About the Author

Joe Scarborough can be seen nightly as the host of *Scarborough Country* on MSNBC. He served as a member of Congress from 1994 to 2001, and resides in Pensacola, Florida, with his wife and three children.

ROME WASN'T BURNT *in a* DAY

ROME WASN'T BURNT *in a* DAY

The Real Deal on
How Politicians, Bureaucrats,
and Other Washington Barbarians
Are Bankrupting America

JOE SCARBOROUGH

Harper
An Imprint of HarperCollins*Publishers*

A hardcover edition of this book was published in 2004 by HarperCollins Publishers.

HarperCollins books may be purchased for educational, business, or sales promotional use. For information please write: Special Markets Department, HarperCollins Publishers, 10 East 53rd Street, New York, NY 10022.

FIRST HARPER PAPERBACK PUBLISHED 2005.

Designed by Elliott Beard

Library of Congress Cataloging-in-Publication Data is available upon request.

ISBN 0-06-074984-9

ISBN-10: 0-06-074985-7 (pbk.)
ISBN-13: 978-0-06-074985-9 (pbk.)

05 06 07 08 09 ❖/RRD 10 9 8 7 6 5 4 3 2 1

To Mom and Dad

Acknowledgments

THANKS TO SUSAN for all her love, support, and typing skills during the writing of this book. Thanks also to Joey and Andrew for being patient and keeping your teenage fights down to a minimum while I was writing away in the study. And thanks also to you, Baby Kate, for providing much-needed hug breaks.

Derrick Kitts did not provide hugs, but he, Noah Oppenheim, and Kim and David Stafford provided great research, while Rick Kaplan provided the most precious resource—time. Thanks to Phil Griffin for finding me, Jim Griffin for protecting me, and Mel Berger for guiding me. Rob McMahon was extremely patient with my brutal schedule and my friends on the *Scarborough Country* staff lightened my load while I finished *Rome*. Thanks to Neal Shapiro for hiring me and "getting it." Thanks also to George, Sara, Carol, John, Alan, Nan, and Supermom Melanie.

Thanks finally to my friends in our congressional office who made my years in Washington personally bearable. Much gratitude is owed also to Earl Durden, Don Gaetz, Collier Merrill, the Booths, the Nashes, and of course, Charlie Hilton, who promised to shoot me dead in Congress if I backed down to the big-spending ways of Washington. Either Charlie is a lousy shot or I kept my word.

Finally, thanks to Cooper Yates for talking me through the early stages of this book while sitting on the deck of the *Fishhouse*. You truly were Dogpatch's Renaissance Man.

Contents

ROME WASN'T BURNT *in a* DAY

Introduction

I AM GROWING CYNICAL in my old age.

Ten years after being swept into Congress as a thirty-one-year-old reformer, I have seen leaders of my political party conspire with Washington's biggest spenders to sell out the future of America. Not too long ago you knew who was on which side. Democrats would get elected by promising to launch new federal programs that would cost taxpayers billions of dollars of their hard-earned money. For instance, Hillary Clinton's recent promise to a group of San Francisco business owners that if they elected Democrats to power, people like them would pay higher taxes for "the common good" shows the Democratic Party is holding up its end of the bargain.

Unfortunately, the Republican Party of my youth is nowhere to be found. The once-proud party of Ronald Reagan is doing nothing to hold true to the ideals President Reagan held so dear—smaller government and less spending.

Under Republican leadership in Congress and the White House, the United States is suffering its largest federal deficit ever. Our national debt is rocketing toward $7,500,000,000,000—that's *seven and a half trillion dollars!* Interest rates are once again shooting upward because of reckless Washington spending. Meanwhile, Democrats and their new ideological allies on Capitol Hill are responding to America's growing crisis by voting through trillion-dollar entitlement programs, massive pork-barrel spending projects, and annual congressional pay raises.

A few Republicans are becoming understandably embarrassed. Before Congress rushed out of town for the 2004 summer recess, a small group of mavericks tried to pass modest budget reforms that would have placed spending caps on legislation and allowed the president to single out pork-barrel bills for elimination. Commonsense proposals like these have been staples of the Republican agenda since Ronald Reagan was president, but when young GOP members placed these provisions in a bill appropriately titled "The Family Budget Protection Act of 2004," Republican leaders worked with Democratic Party bosses to crush the reforms by a vote of 326 to 88.

These days when it comes to wasting your tax dollars there's not a dime's worth of difference between Republicans and Democrats. Sadly, less than a month after the world watched Ronald Reagan ride off into the California sunset, the

Wall Street Journal declared Reagan's party spiritually dead.

"Once upon a time, in a Congress far, far away," the *Journal* began in a June 30, 2004, editorial, "Republicans believed in smaller government. But you wouldn't know it now."

The conservative newspaper concluded its "GOP Lost Souls" editorial by suggesting that the Republican Party would soon be cast out of power if it continued betraying American taxpayers.

"Republicans should understand, principal aside, sooner or later they are setting themselves up for a political fall. If Republicans won't campaign against spending to reduce the federal deficit, they can soon expect to find themselves back in the minority."

In a political season where partisans accuse their political opponents of killing little children to gain access to oil pipelines, the *Wall Street Journal*'s criticism of Republican leaders must be disconcerting to demagogues like Michael Moore and Jesse Jackson. These cartoon characters can vilify Republicans by portraying them as baby killers or slaveholders, but they can't teach you a thing about how Washington, D.C., really works.

I can.

What you are about to read isn't the stuff you see in civics books or newspapers. Instead, I'm going to give you the Real Deal on how the White House, Congress, and Washington bureaucrats conspire to ensure their political survival while sticking American taxpayers with the bill. If you are an avid reader of political books, you probably expect me to explain this nasty little scheme by launching a blistering attack on one political party while lavishing rapturous praise on the other. After all, isn't that the rage in the publishing world these days?

Republicans make millions writing books that paint Democrats as treacherous, evil little beasts who possess dwarfish hearts, while Democrats get rich slandering Republicans as conniving liars who bend the truth as they send young Americans to their early deaths for a few extra gallons of oil—lying all the while. Oh, yeah, and did I mention, these Republicans seem to lie a lot.

This profitable form of political hate speech reached its ugly climax with the release of Michael Moore's *Fahrenheit 9/11*. What once would have passed for bizarre Internet conspiracy theories suddenly became mainstream cinematic fare and the object of praise by Democratic congressmen, senators, and party bosses.

Was it a dream? Did my eyes really see Democratic leaders praising a film that was so virulently anti-American that the terrorist group Hezbollah offered to help distribute Moore's movie in the Middle East?

It had to be a dream. That could not have really been Democratic National Committee Chairman Terry McAuliffe and half the Democrats in Congress at the *Fahrenheit 9/11* premiere sucking up to Michael Moore. After all, it was the same Michael Moore who opposed the United Nations involvement in Iraq because most Americans supported the war, and "that the majority must now sacrifice their children until enough blood has been let that maybe—just maybe—God and the Iraqi people will forgive us in the end."

Had hatred for George W. Bush really infected the souls of

liberals so deeply that they eagerly embraced Moore as their political savior—despite the fact that he defended Iraqi terrorists killing American troops?

"The Iraqis who have risen up against the occupation are not 'insurgents' or 'terrorists' or 'The Enemy,'" Moore wrote in an April 24, 2004, screed posted on his website. "They are the REVOLUTION, the Minutemen, and their numbers will grow—and they will win."

I can understand why Hezbollah wants Moore's message spread across terror camps in the Middle East. I just can't figure out why the Democratic Party would embrace a man calling for the death of young American troops and the success of Iraqi terrorists.

I wonder how Michael Moore's hate speech plays in the homes of Americans who saw their child's head cut off and shoved in front of a video camera by these "Minutemen." Not well, I would imagine.

As a congressman, newspaper publisher of the *Florida Sun,* and host of MSNBC's *Scarborough Country,* I understand that we live in mean political times. Impeachment, the 2000 election, September 11th, the war in Iraq, and a score of political battles going back to Watergate have hardened even the most pragmatic minds inside the Beltway. But as one who has peaked behind the Wizard's curtain, I can assure you that neither party comes to the table with clean hands. That is why simply picking a political party to cheer for while slamming the other side as the singular cause for Western civilization's decline is not only misguided, it is dangerous. Americans who buy into this political circus act are distracted from the real sickness that infects Washington.

Unlike George Washington or the Roman general Cincinnatus, who left his farm to save Rome and then promptly returned home after victory was secured, most political leaders in Congress seem fixated on holding power in perpetuity by funneling taxpayer funds to their favorite pork-barreled projects. Their worldview is limited to a two-year term and a single congressional district. Making matters worse, each member secures his own election (and re-election) by trading whatever votes are required to pass pork-barrel bills that keep his voters happy. And because Republicans and Democrats conspire to gerrymander one another's districts so incumbents are rarely challenged at the polls, the turnover rate on Capitol Hill is lower than in the old Soviet Politburo. This means reckless politicians rarely have to pay the price for their misdeeds. Who does, you ask? You do.

For readers seeking partisan insults instead of political insights, I can assure you that your local bookstore will provide scores of titles guaranteed to reinforce any preexisting prejudices one might harbor against a wide array of political enemies. But this is not such a book. The goal here is to show you how Washington truly functions by taking you behind the closed doors of Congress, into Oval Office meetings, onto Air Force One, and deep inside the corridors of power to which few Americans are granted access.

Many political experts selling partisan comic books these days haven't actually been behind enemy lines long enough to know what goes on in Washington, let alone tell you who you must hate and why. In fact, most recent and notable propagandists have never spent a single day working alongside Washington power brokers or attending meetings shut off from staff members, the press corps, and the outside world. If they had, their *Pleasantville*, black-and-white version of Washington would have immediately been shattered by the cold realities lurking inside the marble palaces our politicians built for themselves on the muddy banks of the Potomac River. Conspiracy theories forwarded by Michael Moore or those put forth in the *Clinton Chronicles* may grab headlines by churning up partisan hatred, but they only further distort America's perception of Washington.

Fat White Pink Boys Beware

I don't kid myself. I was a politician long enough to know this book will enrage the Washington insiders I have long identified as "Fat White Pink Boys"—a term I picked up while serving in Congress. Anyone who has worked more than a few minutes on Capitol Hill knows of whom I speak.

Fat White Pink Boys are a certain class of political hacks who checked their manhood at Washington's city limits. These suspender-wearing political operatives put the *sick* in *sycophant* and will do anything within their power to protect their bosses, their government jobs, and their standing among peers. And because living the life of a political leech is all most of these Fat White Pink Boys have ever known, they can't comprehend why anyone (such as myself) who has ever drawn a paycheck from the Feds would attack the Mother Ship that nurtures all of their political and financial needs.

The Mother Ship in this case being the federal bureaucracy and all its enablers.

I suspect Republican FWPB will be the most offended by this book. After all, party bosses teach you early on that the golden rule of the Beltway is to protect your own. The political operative who is willing to lie, shred, and cover up for his party is a valuable commodity to Washington heavyweights. This might explain why so few members of Ronald Reagan's party have dared to speak out against the shameless performance of Republicans on federal spending, the budget deficit, and America's nearly $7.5 trillion debt.

Congressional rookies are taught the importance of party loyalty upon their arrival to the nation's capital. Freshmen congressmen and senators are first shown the proverbial carrot. Party leaders let new members know that those who play ball will be rewarded. Loyal soldiers will be assigned to the best committees, powerful leaders and chairmen will be put on their fundraising committees to twist arms in the lobbying community, and maybe, if they are especially well behaved, the Speaker of the House or the majority leader will travel to their home district to help raise a few hundred thousand dollars. Starting one's congressional career with a huge fundraiser is a good way to assure a freshman congressman that his toughest election is behind him.

If the carrot fails to entice, the stick is quickly pulled from the party leader's back pocket to beat the wayward member over the head. And if a member dares to speak out against his party leader's latest stupid bill or embarrassing statement, the offending member is quickly reminded that the party neither forgives nor forgets—ever.

For movie fans, I refer you to the kissing scene between Michael and Fredo Corleone in *The Godfather: Part II,* when Michael lets his older brother know the penalty for going against the family is death. Before having him killed, Michael sounds like the U.S. senator his father always wanted him to be when he says to his disloyal brother, "Fredo, you're nothing to me now. You're not a brother, you're not a friend. I don't want to know you or what you do."

Party leaders, like the mob, give no credit to those who show loyalty to the family only 90 percent of the time. When I left Capitol Hill, a Republican Party official who had shamelessly sucked up to me during my years in Washington told a reporter that he was glad I was finally getting out of Congress because, "We were getting pretty tired of Joe Scarborough always attacking the Republican Party."

Always attacking the Republican Party? That would come as breaking news to Bill and Hillary Clinton, Al Gore, and scores of other Democratic leaders who had to endure my appearances on national news programs hundreds of times during my congressional career. Whether during the Contract with America, the balanced budget debate, the government shutdown, or the 2000 election, I was regularly in lockstep with the Republican Party line. But the GOP leader's reprimand for my being "disloyal" to fellow Republicans underscores a more important truth in Washington politics: The party demands nothing less than blind loyalty to its leaders. This reality is critical for conservatives wanting to understand why well-intentioned Republicans and Democrats have betrayed their core principles, broken campaign promises, and bankrupted America.

For most Republicans and a handful of Democrats, the pathway to Congress was paved with campaign flyers promising *Less spending! Less taxes! Less regulation!* The ideological truth behind this Reaganesque campaign cliché is the belief that as the government's power increases, the individual's freedom decreases. This equation is simple. A centralized state's growth is fed by the money and personal freedoms of individual Americans. The more you are taxed, the more money the federal government accrues. The more money the federal government has, the more power it has. The more power it has, the less power you have. It is a simple concept that Republicans once grasped.

But these days Republicans are in no mood to be reminded that they seized control of Congress in 1994 by promising voters to cut the size of government, balance the budget, and return power to individual Americans.

Sadly, the first draft of history has painted these Republicans as liars.

I am sure that Democratic hacks will also be annoyed that I tell the truth about their conspiracy with Republicans to bilk billions from American taxpayers. Democratic bosses play tough for the cameras. They even embrace "documentary" filmmakers who seem to cheer for Islamic terrorists while supporting the killing of American troops for sick redemptive purposes. Democratic politicians must think this hate speech impresses their political base.

But when the cameras are turned off, these same Democrats buddy up to their Republican friends in Congress because, in the end, all Washington politicians are on the same team. Victory for this team is achieved when all members get enough pork siphoned to their districts to ensure re-election and a splendid time for all. Meanwhile, federal spending and the national debt grow exponentially.

Occasionally a Republican like Dick Cheney takes a Democrat's slander personally, and he'll tell a senator to conduct a certain sexual act upon himself. But in the Capitol corridors, this is the exception rather than the rule. Democrats and Republicans huff and sneer for the camera but are usually on the same side when it's time to spend your money. I know Democratic insiders will dismiss this political truth as ideological drivel from a right-wing television host who was once a right-wing congressman. One competing talk show host has even gone so far to dismiss me as a "former neo-Nazi congressman."

But don't be distracted by such insults. In Congress, I did fight the conservative fight when it came to protecting American taxpayers from big-spending politicians. But I was decidedly more moderate than most of my GOP colleagues when it came to issues like the environment and human rights. I never really figured out what was so conservative about paying royalty relief to big oil companies, or corporate subsidies to mining interests, or giving free passes to loggers in national forests. In fact, I never figured out how taking shortcuts concerning America's environment for a quick buck was conservative. It seems like an obvious contradiction to me.

But Washington politics is fueled by such contradictions. That is why I am writing this book. You need to know what your representative does behind your back when he is Washington. You need to know that he is probably more loyal to his political party than to the needs of your family. You need to know that Washington demands such loyalty and why the congressman you elect chooses the party over the people almost every time.

Regardless of the prevailing wisdom in Washington, I still believe loyalty should attach more to ideas than party platforms. That is why when either Republicans or Democrats screw up, I have considered it my responsibility as a reporter, a congressman, a newspaper publisher, and now an author to tell you the ugly truth. And whether my former colleagues in Washington like it or not, that is exactly what I am about to do.

Chapter One

The Real Deal

TELLING AMERICANS the uncomfortable truth about their government has angered Democratic presidents, Republican Speakers, and party bosses of all stripes for as long as I can recall. But this truth-telling approach I've adopted was responsible for getting an unknown thirty-one-year-old elected to Congress in a Florida district that had not sent a Republican to the House of Representatives in more than a hundred years. And it was that same straight-talking style that got me re-elected in landslide elections three times. But my methods did little to endear me to party bosses or committee chairmen who preferred to pepper news outlets with press releases to keep the natives happy while operating quietly behind closed doors constantly consolidating power. As was the case in Congress, I still believe that sunlight is the best disinfectant to cure the ills of Washington. So political hacks of the world beware. It's time for Americans to tear down the gates of the Imperial Congress and let some sunlight illuminate the grim realities of congressional business. As I say on my cable news show, it's time for the "Real Deal."

My fellow Republicans, Congress, the president, and national party bosses have conspired with Democratic politicians, lawyers, and bureaucrats to indulge in the most reckless federal spending spree in U.S. history. How reckless, you ask? Well, in just three years since George W. Bush was elected president, your Republican-run Congress took a $155 billion surplus and turned it into a staggering $455 billion deficit. These self-described conservatives did it in part by passing a staggering array of pork-barrel bills, billion-dollar farm subsidies, and trillion-dollar entitlement programs that America cannot afford.

At the same time, Republican congressmen voted into power in 1994 under the platform of cutting wasteful Washington spending have worked with liberal Democrats to strap you and your unsuspecting fellow Americans over the course of the last decade with the largest annual budget deficit and federal debt in the history of the United States. Most get re-elected with ease by telling the folks back home that they still support smaller government and balanced budgets. But when they return to Congress, these politicians meekly fall in line. That is why the U.S. National Debt Clock in New York City blinked wildly out of control as it streaked somewhere north of $7 trillion.

Seven Trillion Dollars! Dear God.

Jesus, Drugs, and Debt

$7,000,000,000,000+ is an impossible numerical concept for most Americans to grasp. Since I spent my undergraduate years at the University of Alabama, I adopted former football coach Bear Bryant's skeptical view of numbers. "It's kinda hard to rally a campus around some math class," the Bear opined. I agreed and avoided math like the plague in college, relying instead on the goodness and pity of others when it came time to comprehend basic mathematical concepts. Fortunately, back when I was in Congress, Michigan representative Nick Smith was kind enough to draw diagrams in crayons to help me understand just how hard it would be to pay off America's multitrillion-dollar debt.

Nick told me back then, as I tucked my abacus safely into my office drawer, that if I had earned $1 million every single day from the moment Jesus Christ was born until the year 2000, I would still not have earned enough money to pay off America's debt. And that ominous calculation was drawn up four years before Congress went on one of the wildest spending sprees in U.S. history.

Even with the federal deficit and national debt spiraling out of control in recent years, Democrats and Republicans both began pushing for the passage of a national drug benefit for America's ever-expanding senior population. Politicians on both sides of the aisle spent most of the Medicare drug debate acting like burnt-out '70s rock stars. Armed with an endless supply of your cash, these public prima donnas continued paying for their drug fix even after the accountants and lawyers gave them the grim news that their budget surplus had disappeared into a sea of red ink. But in Washington, the band always plays on. There are no Willie Nelsons, no Mick Fleetwoods, no MC Hammers, no repossessed hot tubs. That's because when Washington politicians run out of money, they just print more! Do I hear someone shouting, "Can't touch this"?

Voting for what America's comptroller general David Walker called an $8 billion boondoggle was indefensible for Nick Smith. As the Medicare drug package approached the House floor, Smith says party leaders came to him with an offer too good to refuse: Support our drug deal and we will funnel $100,000 into your son's congressional campaign. But Smith just said no to drugs, voted against adding $8 billion to America's debt, and caused an ethics investigation to be launched into the alleged bribe.

Even mathematically challenged stooges like myself are able to figure out that Washington politicians are mortgaging our economic future and forcing this generation and the next to ultimately pay a painful financial price.

Forget for a minute the economic apocalypse our country will endure when baby boomers push Social Security, Medicare, and retirement programs to the brink of bankruptcy. Instead, focus on what your congressman and senator are costing you and your family today.

Most Americans work half the year paying off taxes, fees, and regulations thrust on them by big-spending politicians at all levels of government. But these days your congressman's culpability is not limited to tax day. You now pay for their carelessness every time you pay your monthly credit card bill, student loan, car lease, or home mortgage. Just as economists have been predicting for years, Washington's runaway deficits are now causing interest rates to move back up. As investment guru Steven Rattner recently told the *New York Times*, "The question is not whether interest rates will continue to go up. But rather how far and how fast."

Economists have long agreed with Rattner that federal borrowing driven by Washington's debt pushes interest rates higher. There is also compelling evidence to suggest that large projected deficits are as damaging to interest rates as current deficits. And because interest rate hikes drive business investments down, Americans pay more on their mortgage, lose value on their homes, and work in a less stable employment climate.

Remarkably, very few in Washington seem to care—despite the fact that the many politicians running Capitol Hill today seized power with empty campaign promises to make Americans more economically secure by balancing the budget.

Leading up to the Republican takeover of Congress more than a decade ago, Republican leaders were relentless in their call for fiscal discipline. On September 22, 1994, Newt Gingrich spoke on the floor of Congress about the "Contract with America" that Republican candidates from across the country would be signing within the week. Gingrich made it clear that reining in Washington spending and its runaway debt would be the central focus of the Republican Party.

"It is time to force the government to live within its means and to restore accountability to the budget in Washington," proclaimed Gingrich.

The future Speaker went on to assure voters that his Contract with America would guarantee that a Republican Congress would be different.

"A campaign promise is one thing," Gingrich told his fellow House members, "but a signed contract is quite another." Gingrich wrote of the Contract, "For the first time in memory, American citizens have a document they can refer to as a means of holding Congress accountable." A total of 367 Republican candidates signed the Contract with America on September 27, 1994. Many of those individuals are now members of the present reckless and irresponsible Congress despite promising "to transform the way Congress works." The Contract with America went on to state, "[Such a] historic change would be the end of government that is too big, too intrusive, and too easy with the public's money. It can be the beginning of a Congress that respects the values and shares the faith of the American family."

Unless the family these members describe consists of a drunk father stealing money from his daughter's piggy bank to pay for his drug habit, this Congress does not share what most of us consider to be America's "family values."

Republicans, like myself, who signed the Contract promised to face the growing deficit head-on. Ten years later an aging American population faces an economic crisis that only grows more daunting by the day. As the U.S. comptroller general noted recently, big-ticket items like Social Security, Medicare, civilian and military retirement, and health care benefits will cause America's debt to spiral even more wildly out of control in the near future. Today the price tag for America's debt works out to roughly $100,000 for every man, woman, and child in the United States. Twenty years from now, that number may increase ten times.

The General Accounting Office also concluded that Washington's current spending practices will require that taxes be increased by 100 percent once baby boomers settle into retirement. If not, federal spending for Social Security, defense, and all other Washington programs will have to be cut in half. And this grave conclusion by the GAO doesn't even take into consideration the consequences of the $8 billion spending spree our elected representatives went on when they passed the Medicare drug entitlement program. The comptroller general concluded that unless Congress and the president act quickly, the financial meltdown resulting from their current policy path will be devastating to Americans. It is not unimaginable to foresee a day when interest rates are at 20 percent, Social Security and Medicare are slashed in half, children are left untreated to die in hospital parking lots because of Medicaid's collapse, while taxes are raised by 100 percent. How did things even get so bad?

Just ten years ago, it was Republican congressmen and senators relegated to the minority who were scolding Bill Clinton and his Democratic colleagues for irresponsible spending habits. Dick Armey blasted President Clinton from the floor of Congress on October 4, 1994, for "throwing in the towel on deficit spending." A month later Armey and Gingrich would take control of the reigns of Congress along with seventy-three freshmen congressmen who promised to put an end to Washington's corrupt culture. As one of these seventy-three rebels, I promised to launch a revolution of reform and stop the reckless spending practices of Congress that were threatening our economy and mortgaging our children's future.

That we dared to lash out at Washington's elite was not an empty populist rant in 1994. Like now, it was a message suited to the times. When I arrived on Capitol Hill, Democrats had maintained a chokehold on Washington's checkbook for forty years. Voters had kept congressional Republicans in the minority since the Eisenhower administration.

But this all changed a few years into Bill Clinton's first term. The deficit was exploding. Taxes were being raised through the roof. The first lady was leading an effort to socialize one-seventh of the economy. Democratic House Speaker Jim Wright had recently been run out of town on ethics charges that arose from his "selling" large quantities of his House speeches to special-interest groups. And Capitol Hill's most powerful chairman, Dan Rostenkowski, was on his way to jail after being indicted on seventeen counts of embezzlement, conspiracy, and mail fraud. Over three hundred congressmen were ensnarled in a banking scandal that involved check kiting, and America's federal debt was streaking toward what was then a record-setting $5 trillion mark. As my grandmom would have said, running against a Congress like that was as easy as "shooting catfish in a barrel." Or put in the more fashionable words of Bob Dylan, "Revolution was in the air."

But a funny thing happened on the way to that Revolution. . . .

As promised, the barbarians stormed the gleaming gates of Congress. But once inside, these Visigoths began stumbling frantically over one another for the honor of becoming a palace guard—or at the very least a member of a prestigious congressional committee. And while older Republican revolutionaries like Gingrich, Armey, and Senator Connie Mack finally got a hold of the federal government's checkbook in 1995 after ascending to majority status, GOP leaders soon were struck by an attack of collective political amnesia. Slowly but surely, they began forgetting the provisions of their "Contract" with the American voters.

Weeks after his stunning ascension to majority leader in November 1994, Dick Armey suggested that since Republicans were now running Congress, term limits were no longer necessary. As columnist Bob Novak noted, "The House roll call votes on term limits were rigged so that every Republican had a chance to vote for one version of term limits while no version actually received enough votes to pass. Hypocrisy was the watchword."

And just as the new majority walked away from term limits, members started suggesting that keeping Democrats in the minority was more important than keeping campaign promises.

This hypocrisy eventually undermined other Contract provisions involving government spending, committee funding, and tax cuts. It was only after the 1994 House freshmen threatened insurrection in 1997 that Newt Gingrich revived the tax cuts he once called "the crown jewels of the Contract with America."

Soon enough, most Republican leaders turned their backs

on the high ideals of individual empowerment expressed by Jefferson and Reagan, and instead followed the lead of George Orwell's barnyard porkers in *Animal Farm.* Our fearless leaders delivered daily lectures to impressionable freshmen congressmen, telling us that the only way to change the system was to dominate the system. Young members were sternly warned that big-spending Socialist pigs had controlled D.C. for forty years through a corrupt patronage system that was fueled by taxpayer money and purchased votes. And if we weren't careful, these Socialist beasts would once again seize control of the barn.

Time and again during those first heady months of the Republican Revolution, freshmen members would be taken aside and told that any procedural or substantive vote that deviated from the GOP leadership's agenda was the political equivalent of treason and that voting one's conscience instead of the party line would only help return Congress into the hands of big-spending liberals like Pat Schroeder (D-Colo.) and Barney Frank (D-Mass.).

In time, GOP party bosses began buying votes and building a patronage system by spending other people's money—specifically yours and mine. And as in the case of *Animal Farm,* it soon became impossible to tell the difference between the farmers and the pigs. By the end of 1998, congressional spending bills laid waste to any remaining pretense that the Republican-run Congress would keep its campaign promises to fight for a smaller, less intrusive federal government.

By the time Bill Clinton and congressional leaders concluded legislative business in 1998, federal spending was exploding at staggering levels. And all those campaign promises that put Mr. Gingrich and the Republican Party in charge for the first time in more than a generation were tossed in the dustbin of history.

Latter-day apologists for the GOP will quickly dismiss my self-criticism by claiming our Republican Congress balanced the federal budget in 1998. But what they won't tell you is that neither Congress nor Bill Clinton had much to do with saving Washington from drowning in its sea of red ink. Instead, perspective gained from the passing of ten years shows that this great balancing act was fueled more by an explosive American economy than by the actions of Washington insiders.

One of my favorite stories from the Clinton era tells of when the president was forced to interrupt a Las Vegas golf outing with NBA legend Michael Jordan after receiving exceedingly grim news. The president's advisers whispered to him that the federal budget would soon balance itself—even if Clinton and Congress did nothing but play golf for the next six months.

Understandably shaken, the president quickly jumped into action and flew back to the nation's capital to sign the Republican balanced budget plan he had furiously fought against for years.

And sure enough, America's budget soon balanced itself.

Democrats still credit the balancing act to their massive 1993 tax hike, while we Republicans point to the spending cuts rammed down the Clinton administration's throat in 1995. But the Real Deal is that Clinton and the Republican Congress were only able to balance their budget because IRS agents collected trillions of dollars in new revenue from American businesses. This new revenue wasn't generated because of Democratic tax increases or Republican spending cuts; it was due to fraudulent booms in the Internet and telecom sectors that were fueled by Wall Street whores and corrupt accounting giants.

While it may be uncomfortable for a politician to put this admission in a campaign flyer, the fact is, it's the only way a congressman could let his voters in on the dirty little secret of how the "unbalanceable" budget was really balanced. Neither tax hikes nor spending cuts impacted the federal deficit as much as the brutish behavior on Wall Street, and in Washington, during Mr. Clinton's Roaring Nineties before the Internet and telecom bubbles burst.

By the year 2000, politicians of all stripes were furiously patting themselves on the back for bringing common sense back to the federal budget process. The new battle lines, we were told, would no longer be drawn around the fight to balance the budget. Rather, we would fight over how to best spend the massive new surplus that rose in the distance "as far as the eye can see." Government watchdog groups like the Concord Coalition and Citizens Against Government Waste joined a handful of Republican contrarians who warned leaders that these surplus projections could disappear as quickly as they appeared. But mavericks like myself, Steve Largent (R-Okla.), Matt Salmon (R-Ariz.), and Mark Sanford (R-S.C.) were once again dismissed as "chicken littles" or members of the "holy jihad." Use of such inflammatory language to describe our aggressive efforts to hold our leaders accountable extended to the leadership itself, as Gingrich, whose Contract we were fighting to defend, repeatedly labeled us as "militants" in his 1998 book *Lessons Learned the Hard Way.*

Unfortunately, the majority of Republicans and most Democratic congressmen rolled their eyes when we continued talking about the tough choices that still needed to be made. "Relax, fellas," we were repeatedly told. "Sure we waste billions of dollars in every spending bill, but look who's sitting in the White House."

"Yeah, but . . ."

"Don't worry, guys. After Bill Clinton leaves town and the White House is disinfected, George W. will assume his rightful place in the Oval Office and then we'll really start changing the way this place works."

Oink.

Meet the New Boss, Same as the Old Boss

No Republican swept into office in the 1994 Revolution could have imagined the spending orgy his own party would launch in 2001. The party of small government quickly morphed into the party of big pork. Washington doled out massive farm subsidies, including $110 million to one rice-growing Arkansas corporation. Florida sugar farmers also got paid for not planting sugar crops. Alabama peanut farmers received money for not planting peanuts and agricorporations from coast to coast didn't have to live off the fat of the land anymore—just the pork placed on their doorstep by Washington politicians.

There are many explanations for what exactly constitutes pork-barrel spending in Washington, ranging from spending bills that serve narrowly defined special interests to appropriations cash that goes to any congressional district other than your own.

In the 2004 edition of the "Congressional Pig Book," Citizens Against Government Waste tried to boil pork down to a more exact definition. For a spending bill to be defined as pork, CAGW says it must meet one of these seven criteria:

- Requested by only one chamber of Congress
- Not specifically authorized by a committee
- Not competitively awarded
- Not requested by the president
- Greatly exceeds the president's budget request or the previous year's funding
- Not the subject of congressional hearings
- Serves only a local or special interest

Congressional leaders like John McCain would add corporate welfare and Pentagon projects, like the recent Boeing leasing scam, to the list. But more on that later.

Pork-filled farm bills shattered spending records in 2002, and massive omnibus appropriations packages stocked with billions in pork breezed through Congress in 2003 without little meaningful debate. This, despite the fact that the 2003 Omnibus Appropriations Bill totaled just shy of $400 billion and was slammed by John McCain as "the mother of all appropriations bills." Of course, leading Democrats like Senator Robert Byrd of West Virginia did launch vitriolic attacks against the Republican leaders' spending plans, but almost always because Democrats believed their Republican opponents *weren't spending enough money on Washington bureaucracies.* My favorite example of this was when self-proclaimed deficit hawk Senator Kent Conrad (D-N.D.) wrote the Republican Speaker a blistering letter on October 21, 2002, demanding that $2.4 billion be added onto the most expensive corporate farm-relief bill in American history.

Which brings me to an important point:

Few serious political commentators would ever dare suggest that a Democratic-run Congress would spend less money than Republicans. Other than proposing cuts in national security and intelligence operations, Democrats like John Kerry have long proposed larger increases in domestic spending and entitlement programs than their Republican counterparts. But most Democrats don't get elected to the White House or Congress by promising to cut spending and taxes while balancing the budget. The fact that Republicans do make this promise every two years makes their bankrupting of America all the more offensive.

The White House's own numbers best illustrate how shamefully the Party of Reagan has misspent our tax dollars over the last ten years. When comparing its fiscal record to that of the Clinton administration, George W. Bush's White House loses in a landslide. This assessment comes from a former congressman whose animosity for Bill Clinton was so deep that when voters asked why I was running for office I would simply answer "Bill Clinton."

Sixty-two percent of the vote later, I was on my way to Washington in 1994.

Within months I was spending evenings on news programs like *Hardball with Chris Matthews* and *Crossfire* criticizing Bill Clinton's economic irresponsibility. After one particularly animated performance, a White House aide who ran into me on Capitol Hill the next day said, "Hey, Congressman, how about taking it easy on my boss."

I laughed and said, "I'm sure the president doesn't even watch those shows or know my name. If he did, he'd hate my guts, but he doesn't."

Clinton's aide stared right at me and said, "Joe, the president does know your name and he does hate your guts."

Relieved to know the feeling was mutual, I was reminded of my mother's insight that I would be judged by the enemies I kept. In 1995, we had a president and a first lady who Dick Armey and a slew of conservatives blasted as Marxists. And while Armey was known to throw insults as hard as Nolan Ryan fastballs, he was not alone in his belief that the economic outlook of the Clintons owed more to Karl Marx than to Alan Greenspan.

But ten years removed from the Republican assault on Capitol Hill, the depressing truth is that the "Marxist" White House of Bill and Hillary Clinton reined in federal spending a hell of a lot better than Armey and all the conservative critics combined, and especially better than our current president, who followed into office the Arkansas Democrat we all loved to hate. While it is true that Bill Clinton had to be dragged kicking and screaming to budget talks to cut federal spending, the truth is that Republicans reined in wasteful spending more effectively under a Democratic president than they do today under one of their own.

Facts Are Stupid Things

Using the Bush White House's own numbers, the federal government under Bill Clinton grew at an annual rate of 3.4 percent. But over the past four years under George W. Bush and his Republican Congress, the federal government has grown at a staggering rate of 10.4 percent. More damning is the fact that during the federal government's most rapid growth, George Bush never once vetoed a congressional bill. Nor did he force his administration or the Republican Congress to make tough choices when it came to spending your money.

I'm not saying Bill Clinton was a bargain for those of us who wanted to reform Washington. He wasn't. In fact, about the only time Clinton tried to stop congressional spending increases was when they had to do with rebuilding America's military—which had suffered serious cuts since 1992. But on all other spending bills, Clinton and his Democratic allies usually whined about how mean-spirited Republicans wanted to starve young children and throw grandmas out in the cold to finance tax cuts for the rich.

In my mind, those were lies far more troubling than any that surfaced during his impeachment trial.

Still, even the fiercest Clinton critics have to face the cold truth that during his first year in office, Bill Clinton forced Congress to make tough choices on the deficit—even if every move made seemed wrong at the time. As Bob Woodward brilliantly detailed in *The Agenda,* Clinton spent his first weeks in the White House ignoring the worst instincts of his political advisers James Carville and Paul Begala, who wanted to jack up federal spending in the face of exploding deficits. Carville and

Begala were outraged that the White House killed a $15 billion "economic stimulus" package at a time when the deficit was exploding toward $300 billion.

Federal Reserve Chairman Alan Greenspan warned the young president that high deficits and such reckless spending would disrupt world markets and wreck the early stages of an economic recovery that began before the former president George Bush left office. Bill Clinton's solution was a $241 billion tax increase that blew up in the Democratic Party's face a year later and cost them control of Congress for the first time in a generation. But many Republicans such as myself often overlook the fact that while we loathed the historic tax hike, Bill Clinton dared to make a politically tough choice to raise taxes and forego a politically popular stimulus package. Instead, the new president sent forth a budget package almost universally despised—one even he privately dismissed as "a turkey." In his shamelessly self-serving 1,008-page autobiography, Mr. Clinton even admits his 1993 tax on Social Security recipients and gas purchases cost his party control of Congress.

But these days, tough choices such as those are simply not made. Instead, Washington cuts taxes while jacking up spending. And when they run out of money, our elected leaders just print more. You know what you call that in Middle America? Counterfeiting. But in Washington, it passes for responsible legislating.

So if you want tax cuts, you get 'em. If you want the biggest farm corporate-welfare bill ever, no problem. Want to increase defense spending to over $400 billion? Sure. How about an $8 billion Medicare drug bill? We know it could bankrupt the system, but why not? We need those senior citizens voting for us. While we're at it, let's propose a massive, pork-filled energy bill. And why not keep conservatives happy by passing another tax cut? Maybe then they'll even forget their president teamed up with Ted Kennedy to pass the biggest education bureaucracy bill ever—even though Republicans promised to abolish the Department of Education just a few years ago!

Republican Party operatives must be thanking God that their base is too stupid to remember party leaders lecturing them for seventy years that there was no such thing as a free lunch. Because these days in D.C., free lunches are being shoveled out to any special-interest group or targeted voting constituency that asks.

I find myself wondering when VH1 will descend on Capitol Hill to produce a "Where Are They Now" special on the Republican Party I knew during the Clinton era. These days the fearless Class of 1994 has dwindled to one or two budget warriors like John Shadegg (R-Ariz.) or Steve Chabot (R-Ohio). But they are the exception and not the rule. Not so long ago—in 1995, to be exact—Republicans were so hell-bent on balancing the budget that we fought to amend the United States Constitution with a Balanced Budget Amendment.

That amendment breezed through the House but was killed in the Senate after California's Diane Feinstein broke her three-month-old campaign pledge to support the measure. Instead, she lied to the citizens of California and voted no. While budget warriors were outraged, a far greater political crime against humanity was being committed by one of our own.

Senate Republican Appropriations chairman Mark Hatfield had the unmitigated gall to come out publicly in opposition to the amendment that most of us considered the Holy Grail of conservative legislative agenda.

Hatfield found himself in a bitter Republican crossfire on Capitol Hill, as the respected chairman became the target of verbal insults and political slander. Most Republicans I knew wanted Hatfield stripped of his powerful chairmanship for this act of political heresy. But in a private meeting with then–Senate majority leader Bob Dole, Hatfield offered his political head on a platter. Bob Dole refused—gaining our enmity for granting a pardon to the Republican Party's new Benedict Arnold. For Dole, the entire "Republican Revolution" was a noisy distraction from his upcoming 1996 presidential campaign. He wanted peace with the powerful appropriations chairman, and by giving Hatfield a free pass he got it.

I wonder what Mark Hatfield's take on Washington is these days. After all, the same Republicans who tried to destroy his political career for refusing to support a Balanced Budget Amendment have now set the record for deficit spending.

Maybe I'm just old-fashioned, but I think someone owes Mark Hatfield an apology. But I won't hold my breath.

Ten years after taking control of Congress, the Republican Party's governing doctrine has morphed from: "The government that governs least governs best" to "The government that's run by Republicans governs best. End of conversation."

But it wasn't that way in the beginning.

Chapter Two

Barbarians at the Gate

WASHINGTON IS FUELED by a nasty mix of ego, sex, and third-rate finger foods. And while a treatise on the latter two would make for inestimably better reading, my publisher has paid me to tell you why Washington politicians waste trillions of your tax dollars every year. I shall do no less than my duty.

But since I've already brought up the other subjects, let me address them in rapid succession. First, I assure you there is nothing in the globs of matter that pass for hors d'oeuvres at Capitol Hill receptions that would move a grown man to do anything other than discreetly spit the offensive food back into his napkin.

As for sex and big-spending congressmen, I must sadly report that there is also very little to discuss on this subject. While I have been told of legends long ago where buxom blondes were strategically placed to garner legislative support from aging subcommittee chairmen, I must report that I have no firsthand knowledge of sex being traded for public funds.

Such a misspent youth.

But I digress.

Instead of sexual tricks or epicurean delights, it is usually the male ego that drives most political and policy trains in Washington. Notice the use of the word "male." My focusing on the male gender does not suggest that female members of Congress are not horrendous beasts. Many are. But Washington politics is still primarily a man's game. And in the pampered, insular world of Capitol Hill, that means staffers and press hounds are forced to endure Old Bulls jutting their jaws out and heaving their chests about in grand acts of puffery.

Pride may cometh before the fall in the Bible, but in Washington it is the breakfast of champions. And vanity is a meal consumed by Democrats and Republicans alike. In fact, Washington in all its grandeur is so arrogant that it assumes it can read your mind, spend your money, and tell you how to live your life. For the uninitiated, this mind-reading trick is helped along by congressmen, senators, and presidents spending truckloads of money—never their own, of course—on polls, focus groups, and taxpayer-funded mail surveys.

Despite such efforts, Washington is *still* kicked in the rear every ten years or so by Americans going to the polls with the expressed intent of sending the elites a message. That message usually involves Middle America overthrowing Washington's established order.

The result is usually a political earthquake.

Since most citizen revolts catch politicians and pundits by surprise, the spectacle of arrogant elites being jarred from their peaceful slumber can be the finest form of entertainment. By the time most populist tidal waves sweep across the Potomac River, Washington political legends of all shapes and sizes are usually the last ones to learn that the political world they once knew has been reduced to rubble. The most notable political "storm" was created by Ronald Reagan in 1980. Although Reagan's legend had reached Churchillian heights by the time of his death, at six P.M. on election night in 1980, Reagan was dismissed as a dimwitted B-list movie star who was an amiable dunce. Just as California political legend Pat Brown laughed off Reagan's California challenge in 1966 when he ran for governor, Jimmy Carter's White House celebrated Reagan's electoral success in the 1980 Republican primaries—thinking the right-wing stooge would be an easier target than George Herbert Walker Bush.

But as in his 1966 gubernatorial race, Reagan's 1980 landslide victory shocked the political establishment while also helping Republican senate candidates capture Democratic seats in Alabama, Alaska, Florida, Georgia, Idaho, Iowa, New York, North Carolina, Oklahoma, South Dakota, Washington, and Wisconsin.

For pure sadists, these lurid events always produce more laughs when it is a conservative landslide like Reagan's that catches Washington and many of its key players off guard. During such times, Americans can entertain themselves by following the media and political elite's desperate efforts to convince themselves that their liberal worldview is not about to go the way of the chocolate brown leisure suit.

Just a month before Reagan's historic landslide, Harvard law professor Lawrence Tribe busied himself by assuring his liberal followers that the independent candidacy of John Anderson made the "odds of a House election seem high." Tribe told *Atlantic Monthly* readers how "politicians, political scientists, and reporters were all haunted by the specter of an electoral deadlock" that would throw the Reagan-Carter election into the House of Representatives. Tribe breathlessly concluded that under such a scenario even Massachusetts's own Edward M. Kennedy might be elected America's fortieth president.

But Americans had something else in mind. Proving just how clueless Eastern elites like Tribe can be when it comes to sniffing out cultural currents, American voters instead chose a conservative giant whose political philosophy was 180 degrees removed from that of Teddy Kennedy.

Americans watching at home are usually positioned best to catch sight of their least favorite anchor or political pundit making predictions that go up in flames under the scorching glare of studio klieg lights as they receive the real-time election results.

Disconcerting exit polls first cross the anchor's desk. But the left-leaning anchor keeps them to himself and dismisses them as unscientific snapshots of a confused electorate.

Then the first polling places close in Indiana and Kentucky. These states often suggest a conservative trend. But such numbers are also breezed over because of the unusually high percentage of Republican voters and incestuous relationships the anchor secretly suspects are responsible for such vexing election results.

But as results begin flooding in from other flyover states between New York and California, the news anchor's self-assurance fades as he is reduced to little more than chin rubbing and head scratching. Soon enough, he begins babbling incoherently, grunting at projections suggesting that yet another liberal icon has bitten the dust. In the case of Ronald Reagan's 1980 landslide election, ABC News legend Frank Reynolds could even be seen gritting his teeth and looking off camera asking, "What the hell is going on out there?"

It's called representative democracy, but certain East Coast TV anchors and newspaper editors always seemed to view it less charitably when Republican leaders like Ronald Reagan or Newt Gingrich ushered in conservative landslides. The *New York Times Magazine,* for instance, dismissed the GOP landslide of 1994 as the result of angry, uneducated white people.

But newspaper editors and the high priests of network news are not the only members of the chattering class who occasionally misread the political mood of more than 250 million Americans. These days a cottage industry exists comprised of politicians, pollsters, talking heads, reporters, lawyers, consultants, and former campaign managers who make handsome livings offering wildly inaccurate projections on TV. Very few analysts who predicted huge Republican losses in 2000 or 2002 paid for their missed calls because so many pundits make so many predictions on so many different levels. Yet for party leaders (other than the Democratic National Committee's perpetual loser Terry McAuliffe), political survival depends on making the right guess about a political election every once in a great while. And every once in a great while, somebody does get it right.

Ten years ago it was Newt Gingrich and his army of Republican insurgents who successfully rode the tidal wave of America's anger toward Bill Clinton to power. Gingrich spent endless hours reading the polls of Republican wunderkind Frank Luntz and ended up believing he could put Republicans in charge of Congress for the first time since Dwight Eisenhower led America through two terms of peace and prosperity.

Gingrich and Luntz drew up the Contract with America and flew across the country for the better part of a year turning the 1994 midterm election into a national referendum on Bill Clinton, Ted Kennedy, and the rest of Congress. At the same time, Gingrich was pressing aspiring Republican candidates to sign his Contract. By nationalizing local congressional races, Gingrich starting making Republicans from Palm Beach to Seattle believe they could knock off enough Democratic incumbents to take over Congress. To drive his point home, Gingrich and other Republican leaders plastered the wall leading into their national party headquarters with a giant sign that read, THINK MAJORITY.

The concept was laughable. Congressional Republicans had managed to be on the losing side of every political firestorm over the past forty years, from McCarthyism to Vietnam to Watergate. In addition, candidates like myself were all the proof anyone in the national press needed to show that the Republican Party of Newt Gingrich was not ready for prime time. After all, Republican voters across America kept electing right-wing candidates who were extreme on spending, taxes, guns, federalism, and abortion.

This crop of candidates who rose from the plains of Middle America back in 1994 were viewed by Georgetown elites as children of the corn—or worse yet, barbarians slouching toward the golden gates of Eastern liberalism's Imperial City. The capital's gentrified elite could only gasp at the odd assortment of freaks, rednecks, religious zealots, Perotistas, hayseeds, Buchananites, intense young party operatives, and scores of other unwashed Visigoths positioning themselves for the final assault. Were these really the rubes America was sending to Washington?

In 1994, Rick Berke, then the *New York Times* Washington bureau chief, visited Northwest Florida to profile my race, and he apparently left deeply disturbed. In an article for the "Week in Review" section that October, Berke dismissed me as a conservative, knuckle-dragging one-trick pony.

"What does Scarborough think of Bill Clinton's medical plan? 'Not the federal government's business,'" Scarborough responded. What about expanding the federal government's role in education? 'Washington should stay out of the education business.' And what about Bill Clinton's crime bill? 'Controlling crime on the streets is best handled by local sheriffs and law-enforcement officers.'"

Rick was polite and his coverage was fair enough, but his story's clear implication was that I was a right-wing whack job that would be buried alive in a political landslide once the good people of Northwest Florida stopped sniffing ether long enough to read my position papers. Most galling to many reporters in 1994 was the fact that so many barbarians running for office, like myself, were political neophytes who had never even been elected county dogcatcher before deciding to run for the United States Congress.

That was a fact that my family found at least as disturbing as the *New York Times'* former Washington bureau chief.

I had decided my first foray into politics would involve unseating an entrenched Democratic congressman named Earl Hutto, who was chairman of the Armed Services Military Construction Committee—a post critical for a congressional district once described as a giant aircraft carrier. Hutto was a sixteen-year veteran of Congress with a daunting built-in advantage—no Republican had been elected in my district since Reconstruction in 1872—and I think that candidate was later hanged.

The fact that my résumé was void of any political experience was also a problem. Before 1994, not only had I never run for political office, I had never worked *in* a political office. Worse yet, I didn't know anyone in elected office, didn't have a dime in my pockets, didn't have a political base, didn't have any name recognition within my district, and I didn't have any idea how I was going to beat a powerful incumbent congressman. That may explain why my parents were less than excited at my new career plans.

"Mom, Dad . . . I've decided to run for Congress. I'm going to run against Earl Hutto because I think Clinton is . . ."

"What?" my mother whispered.

"I'm running for Congress against Earl Hutto," I continued. "I really believe that now is the time for our country to *blah, blah, blah, blah . . .*"

My parents stopped listening and instead focused intently

on whatever was resting on the tops of their shoes. When I finished, my father slowly lifted his head and shifted his focus from the ground squarely into my eyes.

"That's fine, Joey," Dad began. "But I'm voting for Earl."

I don't remember exactly what happened next, but I do remember picking myself up off the floor and turning to my mother for a lifeline.

"Hey, Mom. A little help over here, please!"

A minute or two later, my mother finally said, "I guess I'll vote for you."

Family insults aside, I took it as a positive sign that I had split the first household I visited in my fledgling campaign. Soon I was knocking on all the doors in my parents' neighborhood and across the First Congressional District of Florida. Eighteen months later, after 10,000 handshakes, 5,000 yard signs, 1,000 speeches, 213 fish fries, 97 debates, 2 Brooks Brothers suits, and 1 red power tie, I was on my way to Washington.

For years after that first election, aspiring politicians would come to my office in Washington asking how an unknown thirty-year-old started a congressional campaign with no money, no support, and no hope of winning—but somehow did manage to win. They never seemed to care for my answer.

"Here's the secret," I whispered as they leaned across my desk.

"Get up earlier than everyone. Go to sleep later. Work harder in between, and spend your four hours asleep dreaming up new ways to win votes." After a long silence, the candidate-to-be would lean back in his chair and say nothing. I knew I'd never hear from him again. A few years later, golfing great Ben Hogan passed away. Friends would remember weekend golfers walking up to the links legend asking him the secret of his success. Hogan would shoot them a glare and simply say, "The secret's in the dirt." Politics, like golfing, requires hard work and thick skin.

Hard work on the ground still wins political campaigns on every level. And for years, Republican leaders credited themselves with knocking Democrats out of power in 1994. Truth be known though, Newt Gingrich, Joe Scarborough, and seventy-three challengers had help from a president who seemed overmatched by his new job.

In 1994, Newt Gingrich wasn't the person most Republican candidates had their eyes on. Instead it was the new scandal-prone president, Bill Clinton. Even political novices instinctively seized upon the fact that the "Comeback Kid" had become the third rail of American politics: If you touched him, you died.

As someone who watched Bill Clinton's every move in the late 1990s as he raised himself from the political dead time and time again, I find it hard to recall just how unpopular the president was two years into his first term. The reasons for his unpopularity have filled books longer than mine, but at the top of the list in 1994 were his huge tax increase, his attempt to federalize medicine by socializing one-seventh of the economy, his refusal to reform a welfare system that told young mothers they could get paid for having babies so long as they didn't marry the father or get a job, his efforts to restrict gun ownership, his handling of military flare-ups at home and overseas, his $200 LAX haircut, his shabby treatment of White House travel staff, and thousands of other issues big and small, real and imagined.

I knocked on thousands of doors and asked voters across Northwest Florida to sign petition cards to put my name on the ballot all the while using Bill Clinton as my political pin cushion. After all, I couldn't afford the $10,000 filing fee that Florida requires congressional candidates to pay. I was a bit shaky when the first few doors flew open, and my confidence wasn't helped by a voter who asked why a kid like myself was running for Congress.

"Um . . . I . . . think it's . . . ah . . . time for a change in . . . Washington, and, uh . . ."

As the door of the fourth home I had visited was about to slam in my face, I muttered the word "Clinton." The door flew open again and the newly engaged Scarborough supporter said, "I hate that bastard. Gimme your card." Just like that, a new political marketing strategy was born.

This visceral dislike of Bill Clinton was not only reserved to parts of the Florida Panhandle affectionately referred to as the "Redneck Riviera." Across America, scores of first-time candidates were shocking their families and friends by throwing caution to the wind and taking on entrenched incumbents. The singular sin of these congressmen seemed to be their support of the new Democratic president. After talking to the seventy-three GOP freshmen who arrived on the Hill along with me, I was surprised to learn how almost all of us were middle-class stiffs with no political lineage, no trust accounts, no Ivy League degrees, and no real chance of winning a seat in Congress. But we did.

My roommate in Congress, Steve Largent, was such a

novice to politics that at one of his first rallies he noticed a political sign that was being waved on stage and asked a party leader what the initials "G-O-P" stood for. His fellow Oklahoma congressman Tom Coburn recalls drawing nothing but blank stares and silence when he told friends he had decided to leave his medical practice to run for Congress. Dr. Coburn's daughter called him from college to tell him he had lost his mind. And sixty-eight-year-old Jack Metcalf from rural Washington State likewise decided late in life to take a swing at big-time politics. The risk paid off and soon Metcalf, Coburn, Largent, and scores of other Republican candidates were living the political version of Walter Middy's secret life. We were all newly sworn-in members of the United States Congress.

I remember a striking conversation I had with Metcalf on the day we arrived in Congress. Here was a regular guy, twice my age, living three thousand miles from Pensacola, Florida, and yet we had arrived in the Capitol sharing a political worldview and an ideological synchronicity that no political party could have orchestrated through blast faxes or talking points.

Fifteen minutes into our conversation, I realized that we had nothing in common and yet everything in common. Metcalf quietly asked if I had read the blizzard of faxes Newt and the Republican National Committee had sent out to us every day.

"No way," I sniffed in a very rebel-without-a-clue tone that was the rage in Washington during those heady days.

"Yeah. I threw most of the stuff away too," Metcalf chuckled.

But the fact that most of us walked so closely in step with our other classmates led me to believe that something was happening in American voting booths that no political party or focus group could measure. To me, it was not a coincidence that all these political rookies showed up in Washington one day to fight to balance the budget, reform welfare, and change Washington's corrupt culture.

In the eyes of entrenched politicians on both sides of the aisle, it was as if we stepped out of an *X Files* episode where aliens took the form of untested politicos set on wreaking havoc on Capitol Hill. Washington veterans could not believe a class of freshmen congressmen would dare exert political influence so early in their career or aggressively take aim at leaders of both parties. With seventy-three Republican freshmen voting in block more often than not, it was usually impossible for Republicans or Democrats to pass a piece of legislation without our support. When one particularly cranky freshman deficit hawk from Wisconsin named Mark Neumann started using his seat on the Appropriations Committee to block spending bills he considered wasteful, party bosses were not amused. They eventually booted Neumann from the powerful committee. But when news got out, freshmen leaders let Gingrich and his lieutenants know they would not pass another spending bill until the Wisconsin freshman regained his committee seat. In the first of many humiliating retreats, GOP leaders were forced to acquiesce to the demands of us "militants" and welcome Neumann back with open arms.

Washington's power players were offended that our group of outsiders didn't know its place. But Republican leaders kept quiet for a while because other than the occasional appropriations flare-up, most of the bombs we were throwing in 1995 were aimed at the other end of Pennsylvania Avenue—

ironically at the politician most responsible for our election.

At our first freshman meeting, Bill Clinton was almost an afterthought. Now that we had swept into office, the president would find himself increasingly irrelevant to a political process that would be controlled from the Halls of Congress instead of the West Wing of the White House. And in the House of Representatives, the freshman class would be on the front line in the battle to reform Congress. Unfortunately, our initial freshman meeting was as chaotic as it was ideological—and would be the first hint of problems ahead for a gang that couldn't shoot straight. The seventy-three new members had seventy-three agendas that complemented each of the seventy-three theories our class had on why Americans had swept seventy-three freshmen to power.

Unfortunately, each of us would usually talk for about seventy-three minutes straight—give or take an hour—when given the chance.

We wanted to abolish the Department of Education, the Department of Energy, the Commerce Department, and the Department for Housing and Urban Development. Some even wanted to get the United Nations out of the United States, tear down the building, and salt the earth to ensure no diplomat could ever live there again. I supported all such proposals and more just for the sake of consistency. My dedication to the cause was evident by the unusual patience I showed sitting hour after hour as my new congressional classmates preached reform, revolution, and radicalism.

At the beginning of the fourth hour of that first meeting held at the stodgy Capitol Hill Club, former Hollywood icon and newly elected congressman Sonny Bono took to his feet at the back of the room. Since this would be the entertainment legend's first words before our historic freshman class, we seventy-two other revolutionaries sat quietly in rapt attention. What would Sonny say? Would this former rock star turn out to be our Streisand? Our Sarandon? Our Sean Penn?

Not a chance.

"Hey, fellas," Sonny began as his head bobbed around, "I promised my wife I was gonna take her shopping today. And if we don't wrap it up soon, she's gonna get pissed off."

A motion was heard from the floor, a quick second followed, and Sonny Bono was out the door faster than you could say "Nordstrom." The hard-core few who voted against adjournment stood around in a daze talking about how this type of behavior was simply not acceptable in our new era of political reform and personal responsibility. This Republican Congress was going to have to work for years around the clock just to undo the harm Bill Clinton had brought about in less than two years in the Oval Office. Bill Clinton was the enemy and he could only be stopped through hard work and unceasing prayer.

Ironically one of the first events a freshman congressman attends is a White House reception where the president welcomes new members to town. I was someone who ran on a platform against Bill Clinton for more than a year, and the very idea of smiling and shaking hands with him was unthinkable. As I told my staff members, "I can't even look at the man's face on TV. How am I going to stand in the same room with him?"

"Deal with it," chief of staff Rachel Sanders said with a glare.

And so I did, as I walked through the gates of the White

House and into the grand East Room, where presidents hold state dinners and formal press conferences. Bill Clinton was anything but formal on this night I had dreaded for days. The embattled president, who was facing a torrent of criticism from all quarters for the Democratic Party's electoral collapse, was in rare form. With his left hand clutching the shoulder of one member while his right arm was wound tightly around another's back, Clinton was at the top of his game and acting as if he had just been re-elected president with 100 percent of the vote.

I approached the president determined to mutter a polite "hello" and then get out as quickly as possible. Moving ever so slowly toward Clinton, I whispered to Arizona congressman Matt Salmon, "I think I'm going to get si—"

"Hey, it's Joe Scarborough," the forty-second president said with that patented grin on his face.

I stood frozen in front of the president for a few seconds, thinking only one thought as he rambled on.

"He knows my name. The President of the United States knows my name!"

In a daze I listened as my new buddy, Bill, went on and on about how much he loved my district, my hometown of Pensacola, how he hadn't been to Pensacola since his senior trip in high school, and how he sure would love to get back there again sometime soon.

"Would ya?" I cried involuntarily with a dreamy look in my eyes.

The singular object of my hatred for over a year continued talking for a few minutes as scores of other guests moved in and out of his orbit. He would touch, squeeze, joke, and chuckle with all of them. But from my perspective that night, I was the only guy in the room. And wow! He was so tall. At least six-five.

I'm not sure how long I stood there talking, but I was profoundly affected by my first personal contact with Bill Clinton. He was charming, electric, and perfectly built for onetime encounters with voters in the snows of New Hampshire or Iowa. (Clinton's charm was less seductive in subsequent encounters. He also appeared to shrink about an inch between meetings. The last time we met I could have sworn that he was no more than six feet tall.)

After getting home that night and scrubbing myself down in a cold shower with an intensity that would have made Lady Macbeth proud, I pondered the deeper meaning of my orgasmic meeting with Bill Clinton. This was the politician with whom we were about to go to war. We intended to end his political career for good—even if that meant dragging aging Republican senator Bob Dole across the finish line the next fall. But this night, I stared gloomily into the future and did not like what I saw.

It would be Gingrich and Dole sharing the stage with Bill Clinton over the next two years, and while Dole was a war hero with a flair for black humor, Newt Gingrich was the real face of the Republican Revolution. That was bad news.

I had been close to Gingrich on several occasions before my first meeting with Bill Clinton. Gingrich did not move me in the least. In fact, the week before meeting Clinton, I had stood behind the Speaker for almost an hour while he gave a moving speech on the future of something or other. But as I walked away from that close encounter with Newt, the only personal impression I was left with had to do with the mounds of gray hair sticking out of his ears. And as any political analyst worth his weight in salt can tell you, personal charisma trumps ear hair every day of the week.

We were screwed.

Chapter Three

Battling the Imperial Congress

I'VE NEVER BEEN much for frat boys. I'll be the first to admit my disdain for these creatures rises from dreadful bouts with self-hatred and overpowering insecurities. But despite personal prejudices, I gave fraternity life a try during my first months of college. After all, it was the 1980s, Ronald Reagan was in the White House, God was making a comeback, and Duck Head khakis with heavily starched Polo shirts were the official uniform of campus life. Today, sitting on the deck of a seaside beach house wearing nothing but a twenty-year-old swimsuit and some salsa I just spilled on my left leg, I wonder what was going through my mind at eighteen years of age.

I've tried to convince myself that my swingin' young Republican pose was some sort of rebellious social statement—my way of protesting against the radicalism of Abbie Hoffman, Jerry Rubin, and a generation of dope-smoking freaks. I believed then, as now, that '60s radicals damaged American institutions and led to the United States' defeat in Vietnam. Then again, they did produce some sweet music.

Perhaps this conservative stance was my way of being a loyal foot soldier in Reagan's cultural counterrevolution fueled by a nation of Alex P. Keatons. Collectively, we would reverse the wrongs of Haight-Ashbury, Woodstock, and *Love, American Style*. And perhaps my fraternity membership was a sociopolitical statement that said to aging baby boomers, "You tried. You failed. Get out of the way."

Or maybe, just maybe, I saw fraternity life as a way to hook up with really hot babes while getting even with those bastards that had always taunted me from the pages of Ralph Lauren ads.

You know the tools—hanging out in the Hamptons with V-neck sweaters draped across their broad shoulders while fabulously swank models clung tightly to these tanned paragons of prepdom as if they were Adam Ant himself. I was determined to become what I had always hated—and as quickly as possible.

After a week in orientation and seeing a lot of attractive girls with Greek letters sewn onto their tennis visors, I knew it was time to meet my insecurities head-on and pledge a fraternity at the University of Alabama. Tuscaloosa was a long way from Southampton or Martha's Vineyard, but I figured that frat life in central Alabama was the closest I would ever come to inhabiting Ralph Lauren's WASPy wonderland.

Three weeks into my grand fraternity experiment, I had had enough. Twenty days was about how long it took to figure out that I was paying good money to suffer the indignity of having Samson, Alabama, rednecks drop scotch bottles on brick patios from three stories above while yelling, "Hey, queerball, pick dat up!"

The last words I uttered as an official member of the Pi Kappa Phi pledge class of 1981 involved telling senior fraternity brothers where they could go while quickly packing up my pink-and-green Lacoste duffel bag and heading home.

As with Congress, I walked away from my fraternity experience and never looked back. I spent my remaining college days playing in a rock band, hanging out with friends (for free!), and launching a campaign to abolish the fraternity-dominated student government. Ten years later I would lead the charge to abolish the United Nations, the Department of Education, and a dozen or so other cabinet agencies. But my collegiate attempt to de-bone the fraternity-dominated student government had nothing to do with policy. It was simply good, clean, anarchic fun.

Since leaving college, I have quelled my raging insecurities by starting a successful congressional campaign when I was thirty years old, hosting my own national news show at forty, and being surrounded by a beautiful wife, loving children, and a group of friends whose company I still enjoy without paying a monthly fee. My contempt for frat boys lessens with each new personal success.

Who says this is not the age of compassionate conservatism?

This journey from fraternal animus to acceptance was a positive development for my political career as the 2000 presidential election approached. That's because the man I began urging to run for president in 1997 was being dismissed by critics as little more than an arrogant, aging frat boy. Since I had only shared written correspondence with then-governor George Bush, I withheld judgment. But I did not consider the characterization to be a positive development for Bush's campaign, specifically, or the Republican Party in general.

To ease my troubled independent mind I traveled to Austin to meet the future president face-to-face. Our first encounter was brief but encouraging. Bush met me and a group of Florida friends in the Texas governor's mansion in early 1999. As he neared, "W" greeted me like a lifelong friend. Grabbing my arm, he said, "Great to see you, Congressman Joe."

For the next ninety minutes the future president talked about his vision for America, how he wanted to keep the federal government's role in education limited, and how he planned to restore honor and dignity to the White House. When asked about press reports that suggested he might not seek the presidency because his wife and daughters had reservations, the Texas governor smiled and told us how much he loved Laura, Jenna, and Barbara.

"But they ain't running for president," Bush said with a smirk. "I am."

Snorts and laughs erupted across the room. The straight-talkin' Texas governor had us at howdy.

It was a virtuoso performance that convinced me that George W. Bush would be America's next president. Like Bill

Clinton, Bush was a master working small crowds. His personal aura carried a level of sincerity that even Bill Clinton could not fake. Over the next eighteen months, I worked hard on the Bush-Cheney ticket, appearing on cable news shows more than two hundred times on behalf of the campaign. I used my time on TV to whack supporters of John McCain, Al Sharpton, Ralph Nader, Al Gore, and a slew of Florida Democrats. In the final week of the Florida election challenge, my chief of staff, David Stafford, and I organized a rally in my hometown to force our county's Democratic supervisor of elections to count all the absentee military ballots. More than three thousand showed up to hear me launch new attacks on the Florida Supreme Court, Al Gore, and that hotbed of leftist journalists, MSNBC. The Pensacola event was the largest post-election rally held in America, and it had the desired effect. More than one hundred votes were added to George Bush's overall tally in Florida at a moment in history when a few hundred might determine the winner of the presidential election.

After the 2000 election wound down, I got the chance to spend another two hours with the newly elected president on Air Force One during a return flight from Panama City's Tyndall Air Force Base to Washington, D.C. During those two hours sitting alone with the president in his private quarters, I was again struck by his personal style, bluntness, and easygoing manner. It has been said that Anton Chekhov made those around him feel the need to be simpler and less pretentious. If true, Russia's patron saint of literature would have enjoyed hanging out on Air Force One with George Bush, snacking on toasted peanut butter and jelly sandwiches and Fritos.

But by the time Air Force One touched the ground at Andrews Air Force Base, the president's handshake, smile, unwavering devotion to country, and untortured worldview had led me to one unmistakable conclusion: George Walker Bush was a frat boy trapped in a president's clothing.

In fact, if I were charged with the responsibility of starting a superfraternity for the ages, my first two recruiting targets would be George W. Bush and William Jefferson Clinton. Bush would be in charge of pledge recruitment, and Clinton would be set loose on sorority row to set up weekend parties.

But I digress.

Despite prior objections, I believed a frat icon raised in the oil fields of west Texas was the kind of president America needed in times such as these. Like Ronald Reagan before him, George W. Bush has been endlessly attacked as incurious, ideological, and dumb. Of course, Bush 43 has not helped his cause by stumbling through press conferences and bragging about not reading newspapers. Yet his awkward, unpretentious style has been more of a help than a hindrance in his political career. Just as academics and media elites ridiculed Ronald Reagan for possessing a simple and dangerously moralistic view of world affairs, so too have they dismissed George W. Bush. Their contemptuous disregard for his intellect has led them to underestimate the president at every turn. And whether during debates, press conferences, or public-speaking events, Bush has always risen to or exceeded the low expectations placed upon him.

Long before Ronald Reagan's death, the *New York Times Magazine* devoted a cover story to the similarities between these two conservative heroes in a memorable 2003 article called "Reagan's Son." But as Peggy Noonan, Ron Reagan Jr., and others close to President Reagan have suggested, George W. Bush is not Ronald Reagan's ideological son. While their approach to global challenges seems strikingly similar, their records diverge dramatically when the focus turns to domestic issues.

From the time he left the Democratic Party in the 1950s, Ronald Reagan was a devoted student of Adam Smith, Frederick Hayek, and Milton Friedman. Reagan's belief in the power of free enterprise and a limited federal bureaucracy took on almost religious proportions for the former movie star. He knew it was inevitable that free-market forces would replace centralized state planning just as American Democracy would ultimately crush Soviet Communism.

It was Reagan's belief that the Keynesian idea of increasing government spending to fuel economic growth was an elite intellectual's justification for Washington to take even more of Americans' hard-earned money and shovel it into federal bureaucracies. Big government was not the solution for Reagan. Big government was the problem.

But in the Republican-dominated world of Washington today, the opposite appears to be true. Congress has been given free rein to go on a spending spree of historic proportions—think John Maynard Keynes on crack. Two wars, record deficits, two tax cuts, enormous farm subsidies, explosive federal growth, and the most reckless congressional appropriators in American history have led the government to the brink of financial ruin.

So why have George Bush and my Republican friends in Congress colluded with Democrats to create the most irresponsible spending fraternity in American history? The answer depends on whom you are talking to at a certain time. White House officials will blame their Republican fraternity brothers in Congress while quietly telling you—off the record, of course—that it's Republican House appropriators and Senate egomaniacs who are being piggish by spending shocking amounts of tax dollars on farm bills, highway projects, and pork-barreled packages.

But go behind the closed doors of fraternity houses like the Senate and House of Representatives and these members will tell you a completely different story. Most congressional leaders swear their only desire is to remain true to the high ideals of Ronald Reagan. But they claim the Bush White House won't let them and say they are galled by the administration's arrogant, heavy-handed approach.

"When we tell Bush's people that our voters won't put up with this crap forever," one congressman fumed, "all we get is the same answer. 'The president wants it.' That's it. End of argument."

That complaint was repeated time and again by the senators, congressmen, and staff members I spoke with. Arizona congressman John Shadegg was less biting in his criticism of the Bush White House but did issue a warning that if House Speaker Denny Hastert and Majority Leader Tom Delay didn't start standing up to the president, conservative congressmen would have no choice but to begin a policy of confrontation—not unlike what the U.S. House experienced in 1995 and 1996. This prediction was repeated recently by the *Wall Street Journal's* conservative editorial page editors.

But unlike the early days of the Republican Revolution, the number of fraternity brothers and sisters a Reaganite like Shadegg can count on has greatly diminished. When asked what other members from our historic class of 1994 could be relied upon in a time of crisis, Shadegg went silent.

"There are some real fighters left," he said after a long pause. "Uh . . . Pat Tromey stands his ground."

I reminded John that Pat wasn't in our class.

"What about Paul Ryan?"

"1998."

"Todd Alvin?"

"Nope."

Becoming frustrated, Shadegg had a staff member pull the ratings index of Citizens Against Government Waste and found that other than himself, Ohio's Steve Chabot was the only member of the 1994 class of revolutionaries whose economic voting record received an "A" from CAGW. The band of brothers that started with enough members to effectively shut down any corpulent spending bill was now down to two members. Shadegg sadly lamented: "It's changed so much around here since you, Largent, and the gang left. It's back to business as usual."

More troubling for Reagan Republicans is the fact that members of this congressional fraternity have done worse than return to the standard operating procedures of the House of Representatives. They have instead made the national debt more severe, while convincing Republican congressmen to buy into the belief that an attack on one frat member's spending bill is an attack on the entire fraternity itself.

The White House has also pulled out all the stops to make sure politically popular bills on education, farming, and defense passed regardless of the price tag.

Like Shadegg, I know the pressure of getting a call from the White House when the voice on the other line says, "Congressman, the president's credibility is on the line. You know he has to win this vote tomorrow, and you know that your vote will decide whether the president is seen as a success or failure. You can't let him down."

Very few congressmen I have known are immune to this classic White House ploy. Personally, I always wondered what type of public official would subvert their conscience or the needs of their constituents to save a president's political hide. Most politicians collapse under the presidential pressure. Others, like John McCain, Steve Largent, and current South Carolina governor, Mark Sanford, never flinched when the choice was between listening to their inner voice or blindly following whoever was occupying the White House.

I first experienced how D.C. politicians placed their manhood in a blind trust during the Clinton impeachment ordeal. Every morning Democratic congressmen would sit around the House gym complaining about what a "degenerate" and "liar" Bill Clinton was. That grousing would continue in the House dining room, on the floor of Congress, and well into the evening at Capitol Hill restaurants. But every time the microphone was turned on in the House well or in the television gallery or in the Speaker's lobby where a gaggle of reporters was always waiting, these same Democratic congressmen would bravely defend their beloved president against mean-spirited attacks from Republicans. At the time, Republican members expressed shock that any national leader would trade his public reputation for the sake of the president. But in a wicked example of political irony, just two years later the tables turned. While it didn't involve a stained blue dress or misused cigars, Republican members were being told to compromise their beliefs on behalf of a bigger cause. That cause, as it always is in the White House, was to help re-elect the president in 2004. Most Republicans complied without the slightest protest.

Why do most representatives usually cave in to the White House's wishes? The answer is simple: The political pressure and long-term repercussions are always perceived by members as being too great to overcome. When party bosses say "play or pay," most weak-handed members play. And since the most rebellious Republican members in Congress honored their term-limits pledge, few remain who have the political guts to say no to their president and other party bosses.

Their collective cowardice has caused Washington to be devoured by a $7 trillion debt.

It is all so unnecessary. As Senator John McCain can tell you, it is not really that hard to say no to a president. When a member ignores threats from presidents, House Speakers, and Senate leaders by voting against a spending bill, retribution does not always follow. That is because the wisest Washington leaders know there will always be another political crisis when his vote is needed. At that point, the political traitor once again becomes best friend to the party boss.

While I am more inclined to sympathize with my congressional brethren on most issues, blaming the White House for runaway Washington spending is a cop-out. After all, the

United States Constitution clearly dictates that all federal spending can take place only in the House of Representatives. Fifty-one senators must sign off on the spending plan and the president must ultimately sign the bill, but no money can be spent by the federal government unless the House first approves it. As such, all of the finger pointing toward the other end of Pennsylvania Avenue is a bit disingenuous.

It is also intellectually dishonest for the White House to blame their Republican fraternity brothers and sorority sisters in Congress for the current fiscal crisis. Just as a dime cannot get spent without Congress's sanction, no budget-busting spending bill can become law without first being signed by the president. And while there is always enough blame to go around in Washington, Harry Truman's motto that "the buck stops here" is proper for a president who has yet to veto a single bill. This while the United States government has racked up record deficits and debt.

This story sounds depressingly similar to conservatives who believed one of their own would break Washington of its free-spending ways.

—

Soon after Newt Gingrich and his Contract with America became the most-talked-about subjects by politicos in America, the new House Speaker led his Republican Congress on a legislative blitzkrieg that shook Washington's political establishment. Minutes after being sworn in, Gingrich and his seventy-three freshman disciples began passing more reform measures than any other Congress had attempted in a generation. In my first day as a United States congressman, I voted to pass a radical bill that required Congress to live by the very same laws that they passed. For too long, senators and House members would vote through laws for the rest of America to live by but then write at the end of the legislation, "And by the way, this doesn't apply to us."

We ended that practice immediately after being sworn in to Congress. Gingrich then pushed through reform measures that cut the number of committees and their staff sizes by one-third, and also slashed their budgets by a third.

We also spent the first day voting to hire an independent auditing firm to root out waste, fraud, and abuse in Washington. Gingrich also had us immediately vote to make committee hearings open to the public and to ban proxy voting in committee.

Knowing that reform and ruling by seniority were incompatible with each other, Speaker Gingrich slapped term limits on committee chairmen and most shockingly to a cynical press, on himself. Unlike newly elected majority leader Dick Armey, Newt Gingrich actually believed that his elevation in power did not justify immediate abandonment of the very principles that put him there. Armey, who was quizzed on term limits soon after being elected majority leader, told reporters that since Republicans had now taken over both chambers of Congress, term limits were no longer required.

Talk about foreshadowing! At the time, though, Gingrich had such a firm grip on the levers of Washington power that move Washington that most considered Armey irrelevant. Throughout the spring of 1995, Armey, Dole, and even Bill Clinton were all forced to sing from Newt's songbook. A humbled Bill Clinton came to Congress to deliver his State of the Union address and shocked Democrats and Republicans alike by making the very Newtonian declaration that, "The era of big government is over." Democrats sat in their seats stunned that their president would so flippantly betray the cause of big government while Republicans were equally incredulous. Wasn't this the same Bill Clinton who passed the largest tax increase in history in 1993? Wasn't this the same president who tried to socialize one-seventh of America's economy by federalizing medicine in 1994 and fought GOP efforts to balance the budget in 1995?

While most Republicans sat on their hands and grimaced, I couldn't help but let out a laugh and rise to my feet. Bill Clinton had just offered his formal surrender to Newt's Republican Revolution.

Gingrich dominated the political landscape throughout the spring and summer of 1995 as Congress passed an amazing flurry of reform packages, along with the first balanced budget plan in a generation. He also took the political risk of trying to save Medicare by reforming the thirty-year-old program. On April 7, 1995, Gingrich celebrated his one-hundred-day blitz on Washington by delivering a presidential-style televised address to the American people in which he reported on the new Republican Congress's remarkable progress. A House Speaker delivering a national address like this was unheard of.

A decade later, it is tempting to dismiss the girthsome Speaker's "State of the Congress" address as a self-indulgent overreach, but at the time it made perfect sense. Gingrich was the de facto leader of the United States over the first six months of 1995, and he had the legislation to prove it. University of

Virginia professor Larry Sabato suggested at the time that Gingrich's speech along with the Contract with America was a historical achievement.

"The Contract was the congressional equivalent of a presidential platform. There have only been a few periods of congressional rule where Congress took the lead," he said.

Gingrich's political moves were masterful and his grip on D.C. firm. Soon after his nationwide address, Bill Clinton was reduced to holding a press conference on April 18, 1995, to tell members of the press that he was still relevant.

"I would remind you," the president began in response to a question asking if his voice could be heard over the noisy Republican Revolution, "the Constitution gives me relevance." Clinton again reminded the White House press corps, "The president is relevant here."

That Bill Clinton, let alone any president, would have to lean upon the Constitution as evidence that he was still relevant to the process of governing illustrates just how large a shadow Newt Gingrich cast over Washington in 1995.

Maybe Clinton was still relevant. But as the Republican Revolution roared to summer recess in August, there was little doubt that Newt Gingrich was King of the Hill and master of all he saw in Washington. But he was also the most despised political figure among the press corps since Richard Nixon. Any politician thinking about slipping into self-pity over his poor press coverage need only look at the unprecedented assaults Gingrich took from all corners throughout his tenure as Speaker of the U.S. House of Representatives.

The more success Gingrich achieved throughout 1994 and 1995, the more vitriolic the attacks from media elites. The *New*

York Times described Gingrich's plan to balance the budget, reform welfare, and cut taxes as "a massive sellout to special interests" and as "offenses against common sense." The *Times* went on to denounce the Speaker as "radical" and "inhumane."

Legendary *Washington Post* reporter Mary McGrory, who would befriend me after my efforts to shut down the School of Americas, was less charitable to Gingrich in 1995. She called Gingrich's Contract "foolishness" that brought only "rancor and acrimony."

Brent Bozell, whose Media Research Center has been studying liberal bias in the press for years, concluded that media attacks on Gingrich and his Contract with America were "the most savage, nonstop attacks I have ever seen." Bozell recalled that Gingrich was villianized daily as the architect of a plot to throw children naked on the street, starve grandmothers, and slash Medicare all to pad the pockets of the greedy rich.

The attacks on Gingrich continued unabated up until the time of his premature departure from Washington.

The *Washington Post*'s Thomas Edsall painted Gingrich as a race-baiter simply because he promoted welfare reform in his campaign for Speaker. Not surprisingly, the *Post* somehow forgot to label Bill Clinton as a bigot when he signed Gingrich's welfare reform bill and used it as a platform for his 1996 reelection.

CBS's Bob Schieffer dismissed the speaker as a "lesser man" and blamed him for "shutting down the government" because of a bad seat he was given on Air Force One. The fact Schieffer would say such a thing despite the fact the constitutionally relevant Bill Clinton vetoed the nine appropriations bills that shut down the federal government is revealing. Bob Schieffer's statement betrayed either bias or ignorance.

Longtime Washington reporter and Fox bureau chief Brit Hume also reported what I observed firsthand in my years working among the Capitol press corps.

"I know a lot of reporters in this town," Hume said in a bit of an understatement, "and their attitude toward Newt Gingrich is poison. They hated the guy, there's no doubt about it. They thought he was evil."

In one of the more glaring examples of the media's hysterical treatment of Gingrich, the *Boston Globe*'s Washington bureau chief David Shribman wrote a page-one article on April 25, 1995, suggesting that Oklahoma City terrorist Timothy McVeigh was Newt's protégé. This slanderous attack made more sense when you considered it came just two weeks after Gingrich's triumphant national address. Nothing made the press behave more radically than Newt Gingrich enjoying yet another success.

If any member of the mainstream press even suggests Newt Gingrich received fair treatment from the media, that reporter is either delusional or such an unmitigated liar that you should never trust another word that rolls off their tongue.

Remember that these observations come from a former congressman and a current press hound who worked throughout 1997 and 1998 to remove Newt Gingrich from his Speaker's post. In fact, when the backs of my freshman class were collectively to the wall, I used the distaste the press had for Gingrich to ensure our political survival—but more on that later. In 1995, Gingrich and his Republican comrades had ridden into town, kicked down the gates of the Imperial

Congress, and showed voters that there were still politicians who kept the promises they made. Even if that resolve made critics in the press hate us that much more.

Throughout the August recess of 1995, I stormed across my district to hold town hall meetings in Destin, Pensacola, Milton, Seaside, and Panama City. At every meeting I was greeted like a conquering hero. I went to Washington promising reform and now I stood before my constituents nine months later with nine of the ten provisions in the Contract with America passed through the House. The story was the same for Republican candidates across America. While Newt's negative press coverage created a drag on his approval ratings, the rest of us were riding comfortably in his wake.

Rarely do I feel pity for politicians. They enter public office knowing what a brutal, dehumanizing process politics can be before they shake their first hand. But twice in my public life I've felt great sympathy for politicians. Once was for Hillary Clinton at the height of the impeachment. Regardless of her perceived arrogance and liberal political views, I found her resoluteness in the face of that hellacious personal crisis to be remarkable. I still do not know the source of that strength during that time, but it was remarkable.

Newt Gingrich was the other political figure whose public flogging by the press made me wince. Like Hillary, Newt proved up to the task by smiling in front of cameras while suffering quietly behind closed doors. But unlike the beloved first lady's, Gingrich's fate was sealed by a belligerent press corps before the Georgia congressman was even sworn in as Speaker of the House. Unfortunately for Gingrich, his removal from the Speaker's chair in 1998 was expedited by undisciplined performances in press conferences, blown budget negotiations with Bill Clinton in 1995 and 1996, and backtracking on key issues like tax cuts and federal spending in 1997 and 1998. It was this last offense that turned Gingrich's own army of freshmen insurgents against their once-feared leader.

Going into the fall of his first year in power, Gingrich and his lieutenants were the fear and envy of almost every politician in America—just as Bill Clinton had been after his unlikely election in 1992. But just three years later, Clinton was viewed by most on Capitol Hill with contempt. The press suggested the former Arkansas governor was in over his head. Democrats privately blamed Bill and Hillary for handing Congress to right-wing crazies. Republicans dismissed Clinton as a political bumpkin whose presidency was a mistake helped chauffeured in by Ross Perot. But Bill Clinton, with the help of rogue political operative Dick Morris, was about to set out on a dangerous political path that would either revive his political fortunes or bury his presidency once and for all.

While Republican congressmen were spending the 1995 August recess taking victory laps around their home districts, Clinton and Morris were busy setting political traps intended to derail the Republican Revolution. And since the presidential primary season was just around the corner, the political play clock was running out on the Clinton White House. All that could save Clinton now was a Hail Mary pass.

Clinton and Morris drew up plans for a two-front political war that first had the White House attacking Capitol Hill Republicans before setting their sights on liberal congressional Democrats. These Clinton supporters soon found themselves abandoned in the political battlefield as the next election rolled around. While Morris had President Clinton triangulate Democrats by moving to the right on issues like welfare reform and the balanced budget, he also planned a full-frontal assault on Gingrich's Republican Congress by scaring senior citizens into believing Newt Gingrich wanted to slash Medicare benefits.

Dick Morris got liberal "McGovernicks" like James Carville and Paul Begala out of the Oval Office long enough to convince Clinton that the only way to turn voters against the Republican Congress was to beat them at their own game. Gingrich had spent the previous year pounding the White House into submission by convincing middle-class voters that their way of life was under attack. Gingrich and his Republican posse had also convinced Americans that President Clinton spent his first two years in office figuring out ways to rob voters of their take-home pay through huge tax hikes while separating Americans from their family doctors by socializing medicine.

The message that the middle class was under assault from Washington was a potent political attack that undermined Bill Clinton's presidency. Now that same president was convinced his political survival depended on co-opting that same middle-class mantra.

Over the next few months, the White House, along with the Democrats on Capitol Hill, launched a "Mediscare" campaign aimed at frightening voters into believing that Newt Gingrich wanted to throw grandmothers into the street and take away their Medicare. Bill Clinton spent the last three months of 1995 repeating this same line ad nauseam: "I am not going to let the Republicans cut Medicare to pay for tax cuts for the rich." It became Clinton's singular message for the next six months.

The charge was so shameless and absurd that the first time we heard it during a conference meeting the entire Republican caucus broke out in laughter. The numbers Gingrich proposed for Medicare growth were identical to what Bill Clinton proposed a year before Republicans took control of Congress. But as they did throughout his career, the Washington press corps pushed Clinton around but always let him out of the corner when his survival was in question.

By the fiftieth time Clinton repeated his shameless charge, Republican laughter dissolved into anger. "What a liar!" The one hundredth or so time Clinton repeated the charge, the mighty Republican Revolution was dead. After that, everything we tried to do was linked to Newt Gingrich and Medicare. It was a public-relations nightmare that couldn't be fixed because no one dared go into the boardroom and tell the CEO that it was his own bad press and poor public image that was getting in the way of the bigger cause.

Unlike George W. Bush, Bill Clinton began vetoing every spending bill in sight. In each case, Clinton told Americans how Republican spending cuts were hurting the middle class by endangering the environment, children's schools, and, of course, grandmom's Medicare benefit. Clinton used his veto pen/stamp so aggressively that he even vetoed the Legislative Appropriations Bill, which funds congressional operations. When asked what was offensive about the bill, the president mumbled something about wanting to send a message.

That message was received loud and clear on Capitol Hill. The open contempt Republicans had for Bill Clinton morphed into disbelief, then anger, and then fear. Every spending bill the president vetoed drew cries of protests from constituents and special-interest groups alike who received funding from an agency now on the brink of being shut down. Congressional negotiators got the president to sign a bill that would briefly keep government agencies running at the previous year's funding level. But as soon as that continuing resolution ran out, Bill Clinton and Dick Morris swooped in for the kill. Clinton continued vetoing almost every spending bill that crossed his desk and ultimately allowed the federal government to shut down. Without a two-thirds majority in Congress, Gingrich's GOP was helpless to reopen the government. The president who had been deemed irrelevant just six months earlier began using his bully pulpit and veto power to blame this colossal bureaucratic mess on Newt Gingrich and his radical freshman congressmen.

By November 1995, the plan Clinton and Morris had to shut down the government had become political reality. Now all they had to do was wait. Clinton began wavering early in the process, according to George Stephanopoulos, but Dick Morris, aided by Gingrich's missteps, convinced the president to see the shutdown through to the end. As Stephanopoulos wrote in *All Too Human,* "As much as I would love to think that we were the sole authors of our success, [the Republican leader's] self-inflicted wounds and tactical blunders made the crucial difference."

As Gingrich admitted in his book *Lessons Learned the Hard Way,* his most grievous mistake was telling the media that he was less likely to work with the president to reopen the government because Clinton gave him a bad seat on Air Force One.

While Bill Clinton and Newt Gingrich were fighting for their political lives, I appeared on a number of news shows to answer the question: "When will the government shutdown end?" My answer was always the same: As soon as interest rates went up or Bill Clinton's approval ratings went down.

By early January 1996, Americans were once again beginning to turn against the president. A *USA Today* poll (which the paper never published) showed Bill Clinton's approval ratings diving into the low forties—perhaps proving the government shutdown was finally starting to exact a toll on a White House whose only answers to this national crisis were vetoed bills and glib one-liners.

Soon after reading those poll results, I was summoned to a Republican caucus meeting in the cavernous Cannon House Office Building. Gingrich took control of the meeting early and told House members in no uncertain terms that it was time to reopen the federal government. Enough was enough, the beaten Speaker told us. Bob Dole had abandoned House Republicans for his presidential campaign, the press was breathing down our necks, and now our leader stood before us, desperately asking for our vote.

Gingrich was desperate to resolve this ugly political crisis and was no longer willing to gamble his political future on a showdown strategy whose outcome was still uncertain.

"I have always let you vote your conscience and have never demanded a vote out of loyalty," Gingrich told us. "But today, I need your vote to reopen the federal government to end this deadlock."

As a onetime revolutionary himself, Gingrich understood better than most how hard it would be for his freshmen to cave in to Bill Clinton—especially now that polls suggested we once again had him on the run. But the Speaker was right in one respect. He had never begged for a vote like this. After an amazing one-year legislative run, most believed Newt Gingrich deserved the benefit of the doubt.

He got the vote from us, but it would be the last time. Few recognized it then, but by giving in to Bill Clinton, Newt Gingrich broke the back of the Republican Revolution. Over the next three years the Republican Congress would once again increase the rate of spending and begin chipping away at the very reforms they put in place their first year in power. Still, the impact of the Republican landslide in 1994 meant Clinton would spend the rest of his presidency debating on our home turf. And when you consider Bill Clinton ran on the themes of welfare reform and fiscal responsibility in 1996, it is fair to say that Gingrich's impact continued being felt even after the shutdown ended.

One night in 1998 when I was on MSNBC's *Hardball with Chris Matthews,* Chris asked me if Transportation chairman Bud Shuster's huge Highway Bill would end the Republican Revolution. I laughed and said that with every new bill came bitter talk on the floor about how "this one really kills the revolution for good." But I told Matthews, the "revolution" died the day we reopened the government on Bill Clinton's terms.

—

We had lost that fight, but four years later we were assured the election of George W. Bush would revive our reform efforts. As I was enjoying my toasted peanut butter and jelly sandwich on

Air Force One, President Bush turned the conversation my way.

"What do you think we need to focus on, Congressman Joe?" the president politely asked.

"Well, Mr. President, I think our spending habits have gotten way too reckless in Congress. I know we have a surplus, but that can disappear as quickly as it appeared." I went on to explain that I understood the problems he might have offending powerful GOP allies in Congress by going after their pet spending bills, but I would work behind the scenes to at least expose the pork-barrel projects for what they were. I was sure we could get enough members on both sides of the aisle to win some budget battles with a little help from the White House.

The president, who convinced me he too wanted to cut spending, liked the concept and said such a movement in Congress would be a great help. I moved the conversation to another topic, just assuming he was being polite. But a few minutes later, Bush came back to the subject by saying he wanted to get me together with his budget director.

"Mitch needs some help from Congress on this real bad," the president said of his OMB director Mitch Daniels. "We'll get you guys together."

Ten minutes later White House chief of staff Andy Card came into the Air Force One cabin to have the president sign a handful of documents. Before Card left the room the president looked up at him and said, "Tell Mitch to call Congressman Joe when we get back to Washington. He's going to help us on a project."

Washington and the world look different enough from any plane. But the view from Air Force One is nothing short of spectacular. But as that great American poet Bill Stafford once sang, "All good things must come to an end," and so it was with my flight on Air Force One. Stepping off the plane, I felt gravity's pull more than usual but still had a spring in my step because of the chance I got to discuss world events with the President of the United States for two straight hours. Most important, it seemed that George W. Bush was ready to complete the Republican Revolution Ronald Reagan began in 1980, and that we rejoined fourteen years later. Maybe all that work on the campaign was going to pay off after all.

Three days later I still hadn't heard from Mitch Daniels. I picked up the phone and left a message at his office that I was calling at the president's request.

He never returned my call.

Chapter Four

Capitol Offense

MITCH DANIELS DIDN'T NEED to call me. I would have been pleased if he or the White House would have called any of a number of deficit hawks in or out of government. I'm certain they would have offered the same warning: Washington politicians were wasting money at such an alarming rate that the so-called surplus might never materialize. Congressmen like Paul Ryan, John Shadegg, Steve Largent, and myself were delivering that message in Congress. Daniels would have heard the same dire warnings had he contacted government watchdog groups like CATO, Citizens Against Government Waste, or the Concord Coalition. But he didn't.

And before long, Congress and the White House had blown through a $155 billion surplus.

Come to think of it, Daniels and the White House could have worked with the same first graders the media employed to test those pesky Palm Beach butterfly ballots in the 2000 election. Since those six-year-olds handled that ballot challenge better than the grown-ups who whined about wasting their votes on Pat Buchanan, it is hard to imagine these children would have been more reckless than Congress was with its $155 billion piggy bank.

So how did Congress manage such a historically dubious feat in two years? Washington insiders blame the recession George Bush inherited two months into his term, the economic chaos following September 11th, and the war in Afghanistan in response to the terror attacks.

While these events certainly required additional spending and caused a drain on tax revenue, the mere fact that America faced so many crises at once made Congress's post-9/11 actions all the more deplorable.

While Washington and New York were still burning, congressmen and senators added pork projects for their home districts onto an emergency spending bill aimed at September 11th relief needs. These shameless actions by politicians made Mitch Daniels's blood boil. He fired off a letter to congressional leaders blasting members for using the death of three thousand Americans to make "opportunistic spending sorties masquerading as emergency needs." In other words, elected representatives larded the September 11th relief bill with "emergency" provisions that were neither "emergencies" nor related to Al Qaeda's attacks on America.

The episode proved once again that when it comes to sucking up taxpayer dollars, some politicians will stop at nothing to grab the biggest pile of cash for their political needs.

For those naive enough to believe the September 11th relief act was an isolated incident of national security interests being trumped by pork-barreling, think again. In the year leading up to the second Iraq war, Congress went on a spending spree of near historic proportions. But while they remembered to place thousands of earmarks on scores of bills, our elected representatives still short-changed U.S. troops in Iraq.

As Army Rangers and United States Marines went into combat in hotspots like Fallujah, Najaf, and Nessariyah, too many young Americans did so without Kevlar vests. The situation was so disturbing to some families that they scraped money together at Christmas to buy their son or daughter the vest that Congress and the Pentagon failed to budget. U.S. troops and their families were left wondering why Congress could find ways to spend $2.4 trillion per year, but when it came to protecting those serving in the armed forces there was no money left.

Army reservist Richard Murphy's mother collected $650 in the fall of 2003 to send her son the protective Kevlar plates that would save his life if shot in combat. Mrs. Murphy is a Pennsylvania schoolteacher who somehow managed to find the money for protective army gear when Congress and the Pentagon could not—underscoring the point that something is terribly wrong when it comes to their priorities.

If Reservist Murphy's mother and other military families only knew the half of it, members of Congress themselves would need Kevlar body armor.

While taking care of our soldiers has not always been the top priority of Washington politicians, members have nonetheless always been more than ready to protect their own political backsides by larding spending bills with pork-barreled projects directed to their home districts.

The lengths to which politicians will go to grab your tax dollars and guide them back to their districts are so creative that sometimes I think they should fund a museum to reward the most creative Capitol-offense schemes used to waste our tax dollars. And why not? Over the past year Capitol Hill representatives have spent our money underwriting just about every half-baked idea for a museum that was presented to them by their constituents.

The Museum of Modern Pork

My personal favorite unauthorized earmark for a museum-funding project goes to the Roswell Museum in New Mexico. While one public policy group termed the UFO museum an "unidentified fiscal project," the Roswell Museum had plenty of company. Consider these taxpayer-funded museums from the past year:

- $90,000 for the National Cowgirl Hall of Fame in Fort Worth, Texas
- $350,000 to the Rock and Roll Hall of Fame and Museum in Cleveland, Ohio

- $750,000 for the Baseball Hall of Fame in Cooperstown, New York (as if Major League Baseball players don't make enough to fund their own museum)
- $725,000 to the Please Touch Museum in Philadelphia, Pennsylvania
- $700,000 for the Silver Ring Thing Museum in Sewickley, Pennsylvania
- $300,000 for the Universal Kitchen Design Project, Iowa State University, in Ames, Iowa
- $400,000 for the Pennsylvania Trolley Museum, which is located, of course, in Washington, D.C.
- $450,000 for the Johnny Appleseed Education Center and Museum in Urbana, Ohio
- $250,000 for the Alaska Aviation Heritage Museum in Anchorage

Congress also spent $500,000 of your money renovating the Coca-Cola building in Macon, Georgia, despite the fact that the international corporation rakes in billions of dollars in revenues every year and saw their earnings climb 37 percent last year alone. And since Coke and ice go so well together, our friends in Washington also doled out $2.2 million for the tiny town of North Pole, Alaska. Why North Pole, Alaska, you ask? Because it is in the home state of Republican senator Ted Stevens, Appropriations chairman and Senate leader pro temp.

Other Capitol offenses passed by Congress included millions for cranberries, wine making, and rock-and-roll programs for kids. Also, $2 million was spent to fly Russian bigwigs to such cultural meccas as the Florida Festival Flea Market. There was the $100,000 used to create a World Food Prize in Iowa, and the $500,000 that went to fund the International Coffee Organization. This money was appropriated to help the world coffee industry—and just in time, considering all those Starbucks storefronts that are abandoned and boarded up across America. It is a darn good thing someone in Washington is concerned about sagging levels of coffee consumption throughout the United States and the world.

But just in case you're worried that Washington's fixation with coffee beans and cranberries is the reason our soldiers are without armored vests or Humvees, rest your weary mind. Congress is working overtime to address serious issues concerning the quality of life for the American GI. Why, just last year Congress directed the Pentagon to pay $1.4 million for a dog kennel at Elmendorf Air Force Base in Alaska—again, coincidentally the home state of Senate Appropriation kingpin Ted Stevens.

While the same congressmen and senators have ignored basic safety concerns in Iraq, the overwhelming majority of Washington politicians supported a taxpayer-funded golf program for Florida teenagers. Called the First Tee Program, this classic Capitol offense was touted as a way to "impact the lives of young people around the world by creating affordable and accessible golf facilities primarily for those who have not previously had exposure to the game or its positive values."

So what is the cost of the Florida golf clinic for teenagers? Would you believe $3 million?

I'm sure you're thrilled at working long hours and paying taxes so teenagers in Florida can enjoy the pleasures of a taxpayer-funded round of golf, but your enthusiasm might be dampened by the fact that the money spent on this bizarre program could have purchased at least five thousand Kevlar vests. If you ask me, this would have been a good way to aid the forty-thousand-plus U.S. troops in Iraq who still don't have this necessary piece of equipment. Then again maybe the five thousand vests aren't the safest proposition in the eyes of congressmen. After all, even though these Kevlar vests would surely lower the number of casualties in the war in Iraq, the idea of protecting our troops would not bring home nearly as many votes as such an essential teen golf program.

Unfortunately, Florida teenagers aren't the only species to receive preferential treatment over our American soldiers. Last year alone Congress funneled billions of dollars to their favorite colleges, universities, and congressional districts for research on bugs, bears, baby seals, and hundreds of other topics. Some Capitol offense projects plucked from the line items of congressional spending bills include the following:

- $1 million for the study of DNA in bears
- $7.8 million for the study of Hawaiian sea turtles
- $825,000 for the study of Hawaiian monk seals
- $1 million to study brown tree snakes
- $6 million to study sea lions in Alaska
- $1 million to study the Mormon cricket in Utah
- $90,000 to study fruit flies in Montpellier, France
- $238,000 toward the National Wildlife Turkey Federation
- $750,000 for sea otter research

Imagine Steak and Shrimp and Shrimp and Steak

Apart from French fruit flies and Hawaiian monk seals, good old American cows apparently hold sway with congressmen and senators. This year $900,000 of your tax money went to Mississippi cows for the noble goal of advancing "cattle management in stream crossings." Rebel cows also raked in $225,000 for the always-critical area of grazing research. But Mississippi bovines were not the sole benefactors of cow-based pork: 225,000 in tax dollars also went to pasture research for Utah cows. And cows from Maine to Southern California surely must have rejoiced when they learned that we humans were forking over $15 million to promote dairy programs *overseas*.

While lavishing well-deserved attention on America's underappreciated cow population, congressional appropriators cannily recognized the truth in what Red Lobster marketers have been trying to beat into thick American skulls for years: that nothing really goes better with steak than shrimp. This must be why pork projects funneled to the shrimp industry have totaled more than $60 million since 1985. Just last year, Americans paid $1 million to subsidize seafood harvesting and marketing, nearly $3.8 million to underwrite something called "shrimp research," and $180,000 to study seafood waste. Funding for projects under the sea also included $631,000 to figure out new and imaginative ways to use salmon by developing "alternative salmon products."

It is necessary to tell you that I have nothing against either the study or consumption of cows, shrimp bowels, or alternative uses of Alaskan salmon. What I do find troubling is that billions of dollars flood out of Washington every year to fund projects that can be defined only as corporate welfare.

If the salmon industry wants to harvest and sell more Alaskan salmon, God bless 'em. I saw *The Perfect Storm*. I like fishermen. But let those who stand to profit from government research underwrite the project and then make money off it if their salmon-packed ship comes in to port. Or put another way, why should some guy in Omaha work his butt off all week and pay taxes to subsidize salmon research that will help an Alaskan corporation? Likewise, Alaskan workers shouldn't be forced to pay for bovine research that will only help Nebraska corporations. But they do.

Congress pays off campaign supporters by forcing middle-class taxpayers to foot the bill for projects that should legitimately be paid by Big Oil, Big Tobacco, and Big Business in general. But since these corporations underwrite congressional and presidential campaigns, elected officials help their sugar daddies accumulate greater wealth while making the rest of us pick up their bills.

One of the most egregious examples of corporate welfare involves the farm welfare payments that get doled out to international agricultural conglomerates every year. When I first ran for Congress, I made sure everyone knew that I would vote to pay Alabama peanut farmers for not planting their corps any more than I would support the payment of Florida sugar corporations for not planting sugar. These welfare payments not only cost Americans when they pay the IRS, they also cost consumers at the supermarket. By granting subsidies to cut production, Congress cuts off imports, lowers supplies, and jacks up prices. If Congress stopped paying farmers not to plant their crops, they would be forced to . . . well, plant their crops. Supply would rise and your cost at the grocery store would go down. But instead, this little Socialist scheme costs consumers billions of dollars.

I agree with Bobby Kennedy Jr. that these corporate payoffs create the ultimate ideological irony in this age of self-sufficiency. Our leaders seem to be saying that Socialism is okay for corporations but not for single moms. What's disturbing is that too many congressmen who find the payment of $100 welfare checks to single mothers immoral vote to pay corporate agricultural conglomerations big bucks not to plant their crops. Such irony is often lost on congressional appropriators. Maybe that is why lobbyists conclude all congressmen are willing, ready, and able to be bought off.

These Guys Have Got Nuts

My first term in Congress I was proud to vote for a plan to abolish all farm subsidies within seven years. The Freedom to Farm Act passed in 1995, but it has been made toothless by a series of farm-subsidy bills that have been approved over the last five years—the latest being the 2002 farm bill pork fest. In 1997, it was time to reauthorize a slew of farm bills, so I began getting calls from lobbyists representing those who grew sugar, beets, wheat, citrus, and peanuts. It seems that railing against Alabama peanut farmers had struck a nerve in my home district since I also represented Florida peanut farmers.

Lobbyists representing peanut interests began streaming into my office as the peanut subsidy vote approached. The vote was going to be close, and I was told that without my vote the subsidy program would die.

Good, I thought, but due to my Southern upbringing I bit my tongue.

"Thanks for coming, guys," I said to the group crammed into my office, "but as you probably know, I promised to vote against your program when I was running for office."

The main lobbyist smiled broadly and then dismissed himself as I said good-bye to the rest of the group. A few days later my chief of staff, Bart Roper, walked into the office and tossed an envelope on my desk. The package contained thousands of dollars in checks from the peanut industry along with a note thanking me for taking the time to visit with them.

Being a simple man, I was confused. Hadn't I told the peanut boys I was voting against their bill?

Soon after I arrived on Capitol Hill, most members, staffers, and lobbyists quickly realized I was not a hard guy to figure out. If I said I was going to support your bill, that always meant you could count on a green light going up next to my name when the tally was counted. I didn't sit around like a lot of nimrods waiting to see how leadership was voting or whether the bill would pass. And when I told you, as I did the peanut industry, that I was going to vote against your bill, trust me, I was going to vote against your bill. No bobbing or weaving here. No coy attempts to gain more leverage for my votes. It was a style that owed more to Jim Brown than Barry

Sanders, but this fact was obviously lost on my friends from the peanut lobby.

The day after I voted against the peanut subsidy program, Bart Roper came back in my office.

"I just got a call from the peanut boys and they are ticked off."

"About what?"

"About you voting against their bill, boss," Bart answered with a smile.

"But I told them I was going to . . ."

My chief of staff who was much more experienced in the ways of Washington cut me off.

"I know what you told them, but they gave you a stack of checks. That was supposed to do the trick."

"Bart, are you telling me—"

"Yeah. Their money for your vote. Doesn't matter what you said. It just matters that they gave you money and you didn't hold up your end of the bargain."

I then let out a string of expletives telling my chief of staff and the entire office what those lobbyists could do with their money. I demanded Bart return the checks. He smiled his all-knowing smile, closed the door behind him, and kept the money from the peanut lobbyists safe and secure in the Joe Scarborough for Congress re-election campaign account.

The most enlightening part of the whole episode had nothing to do with the character or judgment of those men and women working for the peanut industry. They were only doing the job they were paid to do. Instead, I was disturbed by the fact that these lobbyists had learned from their years on Capitol Hill that it didn't matter what a member says before the checks roll in. It only matters how the member votes after his campaign account has been stuffed with special-interest money.

Here is one more insight from a guy who has peeked behind the curtain shrouding Congress: Lobbyists are not dumb and they rarely waste their money. They place only the most calculated bets on members and votes they are sure will play out in their favor. The fact that these lobbyists dumped a pile of checks on me after I told them I would vote against their bill, and then called back shocked that I did indeed vote against their bill, leads me to only one conclusion: Their experiences taught them with a high degree of certainty that members of Congress and the Senate could be bought off.

The good news for American taxpayers is that for a few brief years the Class of 1994 was a dangerous bet for lobbyists. The bad news is that the congressional game has tilted the odds back in favor of the House. Washington is once again open to the highest bidder, and the shameful spending bills passed by Congress over the last three years show that it is the American taxpayer who ends up facing longer odds every day.

Chapter Five

Not a Dime's Worth of Difference

I COULD FEEL THE HEAT rise from the base of my neck as the voter in the crowd continued his barrage of leading questions. I was about to let the guy have it, but since this was one of my first congressional town hall meetings, I decided it would probably be best to smile like an idiot instead.

Then he landed a left hook below the belt.

"There ain't a dime's worth of difference between Republicans and Democrats. Hell, your boy Newt's no different than Bill Clinton."

No different than Bill Clinton? Where had he been the last

six months? Albania? Comparing anyone to the almost universally despised Bill Clinton in the spring of 1995 amounted to political slander.

But here in my very own town hall meeting, that is exactly what was being done.

Being a hot-headed punk with a congressional pin, I decided the time had passed to check my temper.

"It's great that you cared enough to come to our town hall meeting," I snarled, "but why don't you read a newspaper before you come to the next one."

As my staff buried their heads in their hands, I continued my counterpunching.

"If you had read any newspaper over the past three months, you'd know the difference between the party fighting for reform and the one who will do whatever it takes to defend the status quo. But since you don't, let me make it real easy for you. Republicans are the ones that pass balanced-budget amendments, Democrats are the ones who have failed to pass a balanced budget in twenty-five years. Republicans are the ones who cut taxes, Democrats are the ones who just passed the largest tax increase ever. Republicans are the ones who vote for measures that make Congress live by the rules they pass. Democrats are the ones who regulate the hell out of Americans and then exempt themselves from the rules they pass."

Being an angry young man was en vogue in 1995 and I continued my diatribe for another few minutes. At the end of it all, I expressed profound shock and regret that any American would be ignorant enough to find little difference between Bill Clinton's Democrats and Joe Scarborough's Party of Reform (It was my party because I had paid for the microphone that day.)

Equating Republicans with Democrats in those heady days of 1995 seemed unimaginable. The GOP was enjoying a historical ascendancy while the Clinton Democrats looked like a tired assortment of '60s survivors, special-interest groups, and coastal elitists. *U.S. News & World Report* even ran a 1995 cover story contemplating the end of the Democratic Party. Bill Clinton's Washington was run by Ted Kennedy (D-Mass.), Pat Schroeder (D-Colo.), Dan Rostenkowski (D-Ill.), and Barney Frank (D-Mass.). For forty years, heroes of the far left had presided over Congress, expanded Washington's bloated bureaucracy, and blocked any political reforms that might make the federal government more responsive. Washington was too liberal, too elitist, and too out of touch with American voters.

But ten years after my first election and the Republican takeover of Congress, I think I owe my grumpy former constituent an apology.

While Republicans under the leadership of Newt Gingrich passed more meaningful legislation during their first few months than any other Congress since LBJ's Great Society or FDR's New Deal, the Republican Revolution quickly ran into a ditch, and every politician began working to do little more than save his own hide.

And that behavior has confirmed the worst suspicions of those who cynically claimed Republicans and Democrats alike were interested in one thing: wasting your money and consolidating their power.

Talk like this always gets me in trouble with my mom, a former Democrat who thought FDR was a king and who lied to my father after she voted for JFK. But she was cured of all Democratic sympathies by 1960s radicalism, violent war protests, political extremism on college campuses, and the social revolutions that brought sex, drugs, and Jefferson Airplane a bit too close to her suburban Atlanta neighborhood. The Southern Baptist FDR Democrat who voted against Richard Nixon would not repeat that mistake twelve years later when Nixon carried forty-nine states to George McGovern's one.

Like many of the Greatest Generation, she now spends her days alternating between grandchildren, the *Drudge Report, Fox News,* and *Scarborough Country*. (MSNBC's decision to hire her son may have persuaded her to turn off Fox News at least one hour a day.)

Many, like my mother, see the fight between Republicans and Democrats as a never-ending battle for the future of America. The unambiguous worldview that paints Republicans as good and Democrats as evil makes it understandable that some conservative viewers are offended when I criticize my own party's failed policies on federal spending, illegal immigration, or troop-level support in Iraq.

I love my mom unconditionally, but neither she nor my Republican friends have explained why it is wrong to hold their own party leaders to the same standard they hold others. Is the conservative Heritage Foundation wrong to draw attention to data that shows Americans are paying more for their federal government in 2004 than at any time since World War II? Should Heritage have hidden those files like Hillary Clinton hid her Rose Law Firm billing records? Should Heritage have ignored Washington's massive discretionary spending growth of 25 percent over the past two years of Republican control? Should thoughtful conservatives duct-tape their mouths and wrap their bodies in cellophane until Democrats once again take control of Congress? Or should they instead warn of an impending storm?

During the 1990s, Republicans rallied behind the libertarian think tank CATO for having the guts to blast Bill Clinton's big-spending ways. But how do you think GOP party bosses feel now that CATO has blasted the Republicans' big-spending ways? They can't be happy since the group has rightfully called George W. Bush one of the biggest-spending presidents in history. CATO has correctly pointed out that total federal expenditures have skyrocketed 29 percent since Bush's inauguration and that his first three years in office have each yielded one of the top five largest annual increases in spending since 1960.

But CATO, the Heritage Foundation, and Republican critics don't place most of the blame at the president's doorstep. Instead, they all rightfully claim that it is the Republican Congress that has failed the American people. As CATO concluded in a 2004 report, "Republicans have clearly forfeited any claim of being the fiscally responsible party in Washington."

In 1995, I could brag to my constituents about GOP plans to eliminate waste, fraud, and abuse in Washington bureaucracies while speaking out on our efforts to abolish outdated federal agencies. But since Republicans took over Congress in 1995, federal funding for agencies of all shapes and sizes has exploded at record clips.

The federal education bureaucracy was a favorite target of Ronald Reagan and then later the 1994 freshman class. The bill I drew up to abolish the federal Department of Education and return the money to local classrooms got 155 cosponsors and was placed in the 1996 House budget resolution. Most Republican activists considered the demise of the educational department and its *Goals* 2000 to be priority number one while Bill Clinton was in the White House. Amazingly since Republicans took over Congress, funding for the Washington bureaucracy has increased by a staggering 101 percent!

The Department of Energy was targeted by Kansas senator Sam Brownback, who told members it was "time to pull the plug on the energy experiment." Far from unplugging the scandal-plagued agency from D.C.'s cash flow, Republicans in Congress have instead increased funding there as well as at all other cabinet-level agencies. For Republicans who can't stand the truth being told about the reckless ways their party spends money, now might be a good time to cover your eyes. Here are the ugly facts about agency-funding increases since the Grand Old Party took over Congress in 1995:

- Justice Department—up 131 percent
- Department of Education—up 101 percent
- Commerce Department—up 82 percent
- Department of Health and Human Services—up 81 percent
- State Department—up 80 percent
- Department of Transportation—up 65 percent
- Housing and Urban Development—up 59 percent

Calling Republican leaders merely hypocritical on the issue of spending somehow falls short of the mark when confronted with these numbers. It is also hard to believe that more Republicans are not following the lead of John McCain, Rush Limbaugh, the *Wall Street Journal*, the Heritage Foundation, and others who have just now begun telling the truth about the Republican Party's sellout of American taxpayers.

If it makes GOP party leaders or White House sycophants feel any better, no one is suggesting John Kerry, Ted Kennedy, or any other elected Democrat short of Georgia senator Zell Miller would spend your tax dollars more wisely. This is because such a suggestion is laughable. Democratic leaders from FDR to Bill Clinton have bragged about their propensity to spend great sums of taxpayers' money in comparison to their Republican opponents. In his recently published autobiography, *My Life*, Clinton described in great detail how bashing "harsh GOP cuts" helped revive his presidency. In July 2004, Hillary Clinton promised California voters she wanted to raise their taxes for "the common good." But this big-government, Democratic approach should not shield Republican presidents or congressmen from criticism when they take a path that could bankrupt America.

Unfortunately in Michael Moore's America, political opponents are painted as corrupt beasts that devolve into the most treacherous cartoon characters. The coarseness of modern-day political dialogue also requires politicians to circle the wagons to protect their tribe while ignoring the failings of their own party. In Michael Moore's America, there is no room for nuance, moral seriousness, or self-examination.

Like many born after the baby boom, I am less interested in waving a political party flag than in getting results. That may be why I changed political parties in the 1980s and 1990s as often as I changed underwear. (Don't ask.) That is also why I was quickly labeled an independent thinker by fellow congressmen and a troublemaker by party leaders when I served in office. This result-oriented approach is why my cable news program goes after Republicans as well as Democrats when appropriate. I never believe the Republican Party talking points any more than I trust what Democratic Party types tell me. I have been to Oz. I have peeked behind the curtain. I can personally assure you that your D.C. wizard is not all he is cracked up to be. He will tell you whatever you want to hear to keep him in his wizard's seat pushing the buttons.

If I sound a little too cynical for your tastes, let me give you some more facts and figures that will reinforce what my town hall voter was talking about so many years ago.

Republicans took over Congress in 1994 in the wake of Ross Perot's dizzying campaign for president. Like Perot, Republican leaders and all of their candidates warned voters about the dangers of the national debt. With Democrats in control of the U.S. House for forty years, the party of Tip O'Neill and Ted Kennedy had allowed America's debt to reach the unheard-of amount of $4.2 trillion.

Americans responded aggressively to our dire warnings and elected Republicans to control Congress—and later the White House. But what happened once the party who advocated less spending came into power? Well, ten years later the federal debt under Republican stewardship exploded to more than $7 trillion. Talk about a failed mission! Washington, we have a problem. . . .

Sorry, Mom. My apologies. But like the man said, there is not a dime's worth of difference between Republicans and Democrats. Instead, there is nearly a $3 trillion difference, and we Republicans are on the short side of that ledger.

For those who argue in vain that deficits, debts, and federal spending always spiral upward in Washington and that the GOP spending explosion was inevitable, recent statistics prove these apologists wrong.

When I was elected to Congress, the National Debt Clock in New York City was steadily blinking numbers indicating a growing debt at a record speed. New York real estate developer Seymour Durst had built the clock near Times Square to show Americans just how badly Washington politicians were wasting taxpayers' money. But six years after my freshman class and I arrived in D.C., Durst's son, Douglas, decided to turn the clock off. The reason was that the federal government was actually running a surplus by the end of my congressional career, causing the debt's numbers on the clock to show a decrease. Instead of piling billions of dollars onto the national debt, the federal government was actually paying off the debt! A strong economy, Republican spending cuts, and a Democratic tax increase all contributed to the federal government running at a $230 billion surplus.

But Durst knew this good news would not last long. On the day he turned off the clock, the real estate developer had the foresight to say, "We'll be ready to turn it back on if things start to turn around. My guess is that politicians will do what they have always done and start spending more money than we can afford."

Watchdogs at the Concord Coalition sounded a similar warning days after Congress began celebrating the news of record-setting surpluses in April 2000. As the Coalition members pointed out, surplus projections were just numbers on a page, and unless Congress and the president practiced disciplined accounting, the rosy scenarios would never materialize.

And the watchdogs were right.

Three years ago the Congressional Budget Office looked at the promising economic landscape and projected a $5.6 trillion surplus for the next ten years under Republican leadership. Today we are faced with a $2.9 trillion deficit that will be added to the national debt over the same time period. And for the first time since 1997, Americans are again faced with deficits "as far as the eye can see," while the Republican president's own budget proposal admits his spending plans will create $1.9 trillion in deficits over the next six years alone. Economists could write long papers on all the reasons why the president's budgets will produce these deficits, but the plain truth is that White House budget staffers decided to spend $1.9 trillion more than the Treasury Department will take in over the next six years. Don't policemen call that check kiting?

To those of us who saw Bill Clinton as a mortal enemy in 1994 and anticipated a Republican Congress being America's salvation, what is most discouraging in the end are these numbers. During Bill Clinton's two terms and a GOP Congress, federal spending grew at a rate of 3.4 percent, whereas government spending has grown at a dangerous 10.4 percent clip during George W. Bush's first term.

I have said that only Congress can legally spend your money, but the Bush administration has done little to interrupt this spending frenzy, especially when one considers that the president still has refused to veto a single spending bill. Beyond sins of omission, Bush's political advisers sent the president out during the 2004 election year to support loopy proposals like a trillion-dollar trip to Mars that probably had less to do with the president burnishing his image as a latter-day JFK than it did with winning a few swing counties along Florida's Space

Coast. And since the president's wife likes the arts, funding for the once-despised National Endowment for the Arts shot to twenty-year highs. About the same time, Britney Spears got married after a fifty-five-hour Las Vegas bender. An understandably concerned White House immediately proposed a $1.2 billion federal program for the government to promote marriage.

While Britney's Vegas fling with her hometown friend probably ended up setting taxpayers back a billion or so dollars, this federalized marriage program was not the first time the Bush White House inched America ever closer to becoming the Nanny State we have long dreaded. As Andrew Sullivan noted in his February 2, 2004, *Time* magazine column, the Bush White House is picking up Bill Clinton's annoying habit of having the feds micromanage social issues that are best left to American community leaders.

Imagine the outcry if Bill Clinton had pushed for a billion-dollar marriage bill during one of his marathon State of the Union addresses. Can't you hear someone (like me) asking the question, "Why does the president seem to trust federal bureaucrats more than preachers, priests, and rabbis? Do they really need the intervention of a federal bureaucracy to advise their parishioners on matters of marriage?"

GOP leaders should be asking those same questions of George W. Bush. But they won't.

But conservative commentator Andrew Sullivan did dare to speak up: "Where once education was essentially the preserve of states, school principals, and parents, this president has expanded the federal role in unprecedented ways. The No Child Left Behind Act holds states and localities accountable for meeting educational standards in order to qualify for federal funds. No wonder Ted Kennedy originally signed on."

Sullivan highlighted other examples of a Nanny State creeping forward in George W. Bush's Washington, including the feds spending millions of dollars in high schools for random drug testing and "character education" for kids.

Maybe you think Washington spending your money to teach our children character education is a harmless enough endeavor. But ask yourself what bureaucracy will determine what passes for good character in American kids? Whose worldview will prevail in a Bush administration? What about in John Kerry's White House? Or in Hillary Clinton's administration? Would the same evangelicals praising George W. Bush's character camps for kids be as pleased when Hillary Clinton's administration promotes radical values that are disguised under the benevolent banners of tolerance and diversity? Wasn't there a good reason conservatives fought Bill Clinton's efforts to nationalize education and social issues during the 1990s? Why are the same self-righteous Republicans who blasted Bill and Hillary Clinton's attempts to nationalize education, history, and values in our schoolrooms mute when Big Government moralism threatens to affect America's educational system being run by one of their own? Do these conservative moralists really believe Republicans will run the White House forever? Or do they simply plan to resume their assaults on the socialization of education when Democrats retake the White House?

I still consider myself an ideological heir to Ronald Reagan. I still believe that the Republican Party of Reagan's day stood for less spending, less regulation, and less government.

And like Reagan's challenge to a Republican incumbent president in 1976, I still believe that patriots have a responsibility to hold their party leaders accountable when they are going in the wrong direction. Despite the grand sendoff GOP party elders gave the Gipper during his state funeral, Ronald Reagan's vision for American government is now a distant memory to most Republican leaders. The ideals he stood for are usually ignored by Republicans until the final, heated days of a political campaign when candidates begin making promises they will soon undoubtedly break.

So what does it all mean? Are Republicans unworthy of public office? Are Democrats more responsible when it comes to managing your tax dollars? Does the mismanagement of your tax dollars by Congress mean that America would be better off with San Francisco's liberal Nancy Pelosi as our next Speaker of the House? Is George W. Bush unworthy of your vote?

The answer to all these questions is not only no, it's *HELL, NO!*

To restate the obvious, Democrats love spending your tax dollars as much or more than Republicans. The only difference is that Democrats usually spend their campaigns flaunting their promise to waste trillions of tax dollars while Republicans hold firm that they will fight those big, bad liberals and end Washington's wasteful ways. Then after the election, the winner—regardless of party—will feed at the public trough while leaving American taxpayers with the bill.

The shabby handling of America's finances over the past four years by the Republican Party means that any GOP candidate claiming his party stands for less government spending and more personal freedom is simply feeding you an election-year line.

Or to put it more bluntly, if a Republican incumbent talks about government spending in a way that makes him sound the least bit Reaganesque, the chances this year are better than even that he is an unrepentant liar.

Chapter Six

Becoming the Imperial Congress

THREE YEARS BEFORE George W. Bush's election to the White House, Republican mavericks still struggled to stay true to ideals that had gotten them elected to Congress in 1994. By the summer of 1997, GOP leaders and the young members who had put them in the majority had little use for one another. With the atmosphere filled with such poisonous air, I walked into Senate majority leader Trent Lott's office to face off on a local military issue. But the conversation quickly turned to national topics. Despite the fact that Lott and I shared the same party affiliation, the same ideological background, and hailed from culturally identical districts, there was little camaraderie by the time I visited his spacious leadership office overlooking the Washington Mall. Florida senator Connie Mack expected my meeting with Lott to go so poorly that he even agreed to sit in as a referee. Mack, though, quickly realized he should have let me walk into this lion's den alone.

"Scarborough," Lott began, "you play music, don't you, boy?"

"Yes sir, I do."

"Well, I play a little music myself. Got a group called the Singing Senators—maybe you've heard of them."

"I have," I said, as my smirk broke into a smile.

"Well, Joe, what would you call it if a drummer was playing off one piece of music, a guitarist was playing something else, the bassist was doing his own thing, and the singer was off singing whatever the heck he wanted to sing? What would you call that?" Lott demanded.

"Jazz?"

Connie Mack slumped in his chair.

"No, boy! You'd have chaos! And that's exactly what you and your friends are causing over in the House going after Newt the way you are. Y'all keep forgetting we're all on the same team."

But by the time of our little meeting in 1997, we all were definitely not on the same team. Leaders like Lott and Gingrich were charged with running the Senate and House chambers efficiently even if that meant busting budget caps by billions of dollars to avoid another government shutdown. Trent Lott did his job as well as anyone, so the Mississippian majority leader kept his position until the White House threw him overboard during a phony race scandal that arose in December 2002. Lott was blasted by the press as a bigot for saying America would have been better off if Strom Thurmond had been elected president in 1948, and his days as Senate leader were numbered. Thurmond, who was celebrating his one-hundredth birthday on the occasion of Lott's speech, had been a champion of racial segregation throughout his 1948 "Dixiecrat" campaign against Harry Truman.

The press immediately seized on the Mississippi senator's unfortunate offhanded remarks, and liberal commentators who had long suggested that Republicans like Lott owed their positions to a racist "Southern strategy" devoured the majority leader in a feeding frenzy. Lott will tell you that frenzy was fed by a White House more interested in playing to the liberal press than showing loyalty to their Senate majority leader.

Not surprisingly, though, similar remarks by Connecticut Democratic senator Chris Dodd a year later praising former Ku Klux Klan member Robert Byrd as a leader who would have served America well at any time during its history were all but ignored by broadcast media. I wonder if African Americans who lived in West Virginia, where Byrd wore a white sheet over his head, would have had as charitable an opinion of Klansman Byrd as did the very white senator from Connecticut? This was, after all, the same ex-Klansman who attacked "white niggers" on national TV a few years back. But as Lott and all conservative white Republicans must know by now, the *New York Times* and the elite Northeast liberal media have a built-in bias against their culture. From *The Dukes of Hazzard* to *Mississippi Burning,* residents of the Deep South are always prime targets for lampooning and villainous caricatures. After the movie *Mississippi Burning* was released, I eagerly awaited the follow-up—"Massachusetts Ablaze"—to take Americans through the tragic days when South Boston schools were integrated, years after those in Mississippi. But of course that movie never came to a theater near you.

But years before the Lott tempest, most maverick Republicans understood the majority leader's job was to whip members like myself into line. Republicans feared the total collapse of the party after the 1996 government shutdown. Democrats again positioned themselves to regain control of Congress, and since it took forty years for Republicans to gain the majority the last time Democrats took control of Congress, Lott and Gingrich were in no mood to tolerate dissent.

At the same time, senior Republican members who had waited decades to take control of the congressional checkbook were ready to enjoy the fruits of their labor. Appropriations chairmen like Hal Rodgers (R-Ken.) and Jerry Lewis (R-Calif.) made no secret that their goal was to bring home as much pork to their districts as possible. Throughout 1995 and 1996, these Old Bulls labored under the cruel reign of Chairman Bob Livingston. During those two years, Louisiana's Livingston was a pork-barreler's worst nightmare. He proposed huge spending cuts and then shocked political hacks by ruthlessly pushing them through Congress and cutting government spending.

The cuts passed by Livingston's Appropriations committee were endorsed by Gingrich and helped America move toward a balanced budget for the first time in a generation. They also proved to the party's biggest skeptics that this Congress would be different. This Congress would keep its promises to

American voters. Perhaps we were operating under a binding Contract after all.

While reformers cheered these historic $56 billion in spending cuts, Democrats and old-line Republicans recoiled at the novel approach to governing that maverick conservatives called "fiscal restraint." Armed with a historical perspective that my freshman class and I lacked, these porkers knew that all they needed to do was wait. Washington always wins, and soon enough their time to wallow in the public till would come since Washington always beats down those who try to reform it.

Of course, they were right.

After the Republican Revolution's collapse during the 1996 government shutdown, big-spending bills once again flooded the halls of Congress with little or no restraint. Gingrich and his entire leadership structure were so dazed by their disastrous showdown with Bill Clinton that no one in Newt's office had the credibility to preach discipline to the Hill's most powerful committee chairmen.

By the end of 1997, Republicans of all rank and ideology began spending with a reckless abandon. As the public-interest group Citizens Against Government Waste grimly noted in 1998, federal pork spending increased more than 50 percent between 1993, when big-spending Democrats controlled Congress, and 1997, when "conservative" Republicans were writing the checks. For congressional observers, this was proof that the great Republican Revolution had failed.

If 1997 was a tactical retreat, then 1998 was an unconditional surrender to big spenders. That was the year House Transportation Chairman Bud Shuster moved his mammoth $217 billion Highway Bill toward passage. And as the beastly bill slouched its way through Congress, one member after another threw on his or her slab of pork. By the time the Highway Bill reached Bill Clinton's desk, it was the most expensive public-works bill in U.S. history.

The Republican Highway Bill included the following:

- $3 million spent on a propaganda film extolling the virtues of highways—as if $217 billion wasn't enough of a ringing endorsement
- $20 million directed at building roads overseas
- $1.5 million to study the packing habits of truckers at their favorite truck stops
- $500,000 to study sidewalks at the Kennedy Center
- $2.75 million to build a smoother access road to a baseball park in Dayton, Ohio

One could list enough ridiculous line items in the 1998 Highway Bill to possibly make the king of all porkers, Robert Byrd (D-W.V.), blush. The late, great *Atlantic Monthly* editor Michael Kelly wrote of the Republican transportation bill in the *National Journal*: "This bill, which is $26 billion over the budget cap agreed to in the 1997 balanced budget deal, is an exercise in vote buying that Tip O'Neill would have been proud to call his own." O'Neill was the legendary liberal Democratic House Speaker who coined the famous—and deadly accurate—phrase "Money is the mother's milk of politics."

Kelly also noted that Shuster's Highway Bill passed the "conservative" Republican House appropriately enough on April Fool's Day. The final tally wasn't even close. Porkers whipped the reformers 337 to 80.

As he would find himself more often than not in years that followed, Steve Largent was on the losing side of this historically bad spending bill. The NFL Hall of Famer said of the Republicans' legislative handiwork, "This bill is everything I ran against. It makes me sick to my stomach."

Connecticut congressman Chris Shays, no right-wing nut himself, called the GOP transportation free-for-all "obscene" and "politics at its worst." Shays concluded, "This is the best indication that the Republican Revolution is dead."

Another indication of just how hypocritical the party of Reagan had become is the fact that Shuster's 1998 Highway Bill was a 250 percent spending increase over a similar transportation package Ronald Reagan vetoed in 1987. Reagan attacked the 1987 pork-filled bill written by the Democrats as too expensive a burden to place on working Americans.

Ten years later Republican leaders who considered themselves heirs to the Reagan Revolution teamed up with most Democrats to pass a bill that would have been vetoed by President Reagan in 1987 or Newt Gingrich before the shutdown.

Few Republicans were as outraged by Shuster's Transportation Bill as Massachusetts' junior senator John Kerry. But Kerry was angered by the fact that Congress had not spent *enough* money in his home state. The future Democratic presidential nominee teamed up with Alabama Republican Dick Shelby to threaten Trent Lott. The two told the Senate majority leader they would block passage of Shuster's Highway Bill unless their states received even more money when the package came to the Senate. According to the publication *Congress Daily,* Lott then "put on his apron and served up his best

$200 million dish of pork"—promising Kerry and Shelby the money would be theirs in a future bill.

The payoff worked, Kerry and Shelby voted yes, and the Transportation Bill breezed through the Senate. John Kerry was to receive his dowry in the bitter, closing days of 1998, by which time Washington's fixation had moved from budget battles to soiled blue dresses. The mean season of Bill Clinton's impeachment had dawned on Capitol Hill.

I came to Congress ready to fight battles over the size and scope of American government, so the impeachment process was a nasty distraction from the issues that mattered most to me. Monicagate also helped distract Americans from more significant Clinton-era scandals involving the funneling of foreign campaign cash to the Democratic Party. Howell Raines, who at that time was the *New York Times*' editorial page editor, characterized the Clinton White House's fundraising scams as the biggest political scandal since Watergate. But the rest of the American media machine yawned while Janet Reno, John Glenn, and a cast of hundreds engaged in a cover-up of Nixonian proportions. Instead of asking the type of tough questions Republican senators once asked of their presidents ("What did the president know and when did he know it?"), Ohio Democratic senator John Glenn told Fred Thompson that uncovering the truth in Chinagate "is your problem." The former White House council and Hollywood movie star shot back, "No, Senator, it is America's problem."

In 1998, a stained blue Gap dress and Oval Office cigars stirred more interest in America's newsrooms than any illegal funding of presidential campaigns. By the time the final remnants of the media circus pulled into town for the Senate impeachment trial, all stories involving Communist campaign donations or the public looting of Americans' tax dollars were relegated to the back pages of America's daily newspapers.

Just as Neil Young reminded us that "Rust Never Sleeps," neither does a politician in search of a quick buck. While the world watched the spectacle of Republicans and Democrats squealing at each other on TV day and night, leaders from both parties were quietly working together to pass yet another record-breaking pork-filled bill.

Since the impeachment drama was little more than a cynical game of high-stakes politics, hard feelings rarely stopped Old Democratic Bulls from working together with Old Republican Bulls—all to complete "the people's business."

What "the people's business" meant at the end of the 1998 congressional session was the most massive omnibus spending bill in U.S. history. The final spending bill that year was 4,000 pages long, weighed 40 pounds, blew through $610 billion, was drafted in less than a week, debated on in a day, and passed without amendment.

The 1998 Omnibus Bill provided 610 billion good reasons for leadership to forbid members from amending the bill. Republican and Democratic appropriators alike had taken pork-barreled spending to a new level. And if reformers like McCain, Largent, Sanford, and Shadegg had been allowed to draft amendments to cut the enormous price tag, little of the spending items would have survived the light of day. But the same GOP leadership that criticized Democratic House Speakers years earlier for passing rules that limited debate and banned amendments to bills were reduced to doing the same thing in 1998. The reason leaders gagged open debate was simple. Items contained in the bill, like the ones listed below, never would have passed Congress as stand-alone bills:

- $6 billion spent in "emergency" relief for farmers
- $1.1 million for manure handling and disposal in Mississippi
- $500,000 for swine-waste management in North Carolina
- $750,000 for grasshopper research in Alaska
- $250,000 to study lettuce genetics in California
- $300,000 for honeybee research in Louisiana
- $175,000 to study desert plants
- $220,000 for blueberry research in Maine
- $5 million to research new and exciting uses for wood
- $5 million in relief for "at risk" fishermen in John Kerry's Massachusetts

The Senate was also very careful to pin their names on just about anything that didn't move. And Americans got stuck paying the bill for the million-dollar vanity name plates.

- $6 million to help start the Robert J. Dole (R-Kan.) Institute for Public Service and Public Policy at the University of Kansas
- $1 million for a Mark Hatfield (R-Ore.) School of Government at Portland State University in Oregon
- $1 million for a Paul Simon (D-Ill.) Public Policy Institute in Illinois
- $1.2 million for the Mitch McConnell (R-Ken.) Conservation Fund in Kentucky

- $12 million for the Patrick Leahy (D-Ver.) War Victims Fund in Vermont
- $1.5 million for the Claiborne Pell (D-R.I.) Institute for International Relations in Rhode Island

So how did Congress bust the budget caps they had set just a year earlier? They simply called their pet projects "emergencies" to get their pork passed. The definition of what constituted an emergency ranged from the ridiculous to the sublime, but my favorite was the "emergency" item that extended the 1998 duck-hunting season in Mississippi.

Since "emergency spending" is not subject to budget caps Congress sets for itself, big spenders pass their pork projects within the legally set limits by abusing the emergency loophole. The Congressional Budget Office reported that Congress spent $154 billion in "emergencies" from 1999 to 2002. And since these bogus emergencies are a big spender's favorite device for busting budget caps, there is no stopgap measure to halt the abusive practice.

Previous "emergency" items documented by Citizens Against Government Waste include:

- $122 million for Russian-U.S. space ventures
- $40 million for NASA's Spacehab module
- $10 million to convert a post office to a train station
- $1.3 million for two sugar mills in Hawaii

The only real emergency in Washington by the end of 1998 did not involve Hawaiian sugar mills or Mississippi ducks, but shameless congressional porkers. After a year of licking their wounds over a failed coup attempt involving House Speaker Newt Gingrich (more about that in a minute), we young Republican mavericks were now ready to strike back at the man who launched the Republican Revolution.

Gingrich had been badly shaken by a budget showdown with Bill Clinton in 1996—and a year later by a Republican coup attempt that almost took him out. Also, by the end of 1998, Gingrich's daily beatings from the Washington press were getting increasingly hostile. The sharks could smell blood in the water.

Even though he ran unsuccessfully as a moderate congressional candidate in 1974 and 1976, Newt Gingrich believed he could deal with members of all ideological stripes. But his tumultuous rise to power through the 1980s and early 1990s cut him off from those angry Democrats he attacked. After the press unfairly painted his Contract with America as right-wing drivel, Republicans representing more moderate "swing" districts also began distancing themselves from their controversial Speaker. Gingrich's political base then consisted of the Class of 1994 and a hundred or so Bob Dornan conservatives—after he cut himself off from Democrats and many moderate Republicans. This situation was unacceptable to Gingrich. If he believed, as former Republican leader Susan Molinari claimed, that Western civilization's future was resting on his weary shoulders, then his governing majority would have to be expanded beyond knuckle-dragging reactionaries and right-wing kooks like myself.

By the time the 1998 Transportation and Omnibus Bills were rocketing through the U.S. House, Gingrich's strategy became clear to Republicans. He would moderate his image by triangulating conservatives and meeting Democrats' spending plans dollar for dollar. These efforts by Gingrich to appease moderate Democrats and make peace with Bill Clinton through bigger spending bills was nonsensical and angered the conservatives who were his only remaining base of support. The Clinton years were not the Era of Good Feeling any more than they were the Age of the Extended Hand under former president George Bush. Gingrich had bitten off Democratic Speaker Jim Wright's hand and sent the once-feared congressman packing for Texas. Michigan representative David Bonior, Wright's protégé, and the rest of the Democratic Party would not forgive or forget what Gingrich did to their House Speaker in 1989. They were waiting to repay the favor when Newt assumed power five years later.

Even though Gingrich should have known from his first days in power that Democrats would not be happy until he was destroyed, the Speaker continued his hapless charm offensive while ignoring friends. Concerned members tried to remind Gingrich that when Reagan got in trouble, he always turned to his base, not his enemies. Frustrated that Newt wasn't getting this message, I took a more dramatic tact.

On a night billed as "The Speaker's Listening Session," a small group of Republican congressmen went into Gingrich's office to discuss ways to improve operations in the House. I decided to make the issue about Newt himself while others were discussing the finer points of political doctrine and parliamentary matters. When my turn came to talk, I told the group about a scene from Oliver Stone's paranoid account of Richard Nixon's life. I recounted a dramatic moment from the film when Pat Nixon got the thirty-seventh president's attention by reaching down and grabbing his coat lapels. Retelling the story, I moved in front of a seated Gingrich and grabbed hold of his jacket while paraphrasing the former first lady: "They'll never love you, Newt. Don't you understand? They'll never love you."

Silence fell over the room. Somewhere in the distance a dog barked. I slowly backed away after doing my best Pat Nixon, while Newt and his aides nervously shot looks at one another. *So much for swinging for the fences,* I thought. I sat down and never said another word.

By the end of 1998, Gingrich was in no mood to be lectured by me or by any other member who he worked so hard to get elected. A love-hate relationship had developed between Gingrich and many in the rebellious Class of 1994. But even Gingrich's fiercest critics would say in quieter times that the Georgia congressman was a gifted orator and strategist whose greatest sin was his loose lips and erratic management style. Members still faulted Gingrich for blaming the government shutdown on his bad seat on Air Force One. Sadly, his immense talents were overshadowed by years of bad press clippings, a nasty case of foot-in-mouth disease, and a continuing failure to stand up to Bill Clinton's big-spending designs. Looking back now, I realize our objections to Newt's handling of the budget seem laughable when you consider the miserable spending records of his successors in Congress.

After years of quietly absorbing criticism from back-bench Republican "militants" like me, Newt Gingrich finally struck back. As the 1998 Omnibus Spending Bill was debated on the floor, Gingrich took a verbal jab at those of us pressuring fellow Republicans to vote no on the final vote of the year. The

Speaker dismissed us as members of "the perfectionist caucus" who lived in a world where we were "petty dictators."

As Oklahoma congressman Tom Colburn later pointed out, the attack was vintage Newt Gingrich. The Speaker shifted the blame for this intraparty fight from Bill Clinton and party elders to those of us struggling to stay true to the words Gingrich spoke to the country during his televised address just a few years earlier.

Gingrich ended up winning the day by pushing the biggest Omnibus Spending Bill through Congress so it could get signed by Bill Clinton. But his angry attack at House Republicans would be the last speech Newt Gingrich would ever make from the floor of Congress.

Soon after we adjourned and the spending orgy was officially over, Republicans were routed in midterm elections and saw their majority reduced in the House to six seats. The day after these shocking results, Arizona congressman Matt Salmon called me with news that would shake official Washington.

"I'm taking Newt out. I've already got enough votes to stop him from being Speaker again."

"You want me to call him?"

"Yeah, if you want to. Tell him it's over and there ain't a damn thing he can do about it."

This day was long in coming for Salmon. Matt was a devout Mormon with a great sense of humor, but he also had a temper when he believed he was being lied to. Matt and many in our ever-dwindling group of mavericks believed our Speaker had ultimately failed us in his fight to reform Washington. By the end of 1998, Congress wasn't just returning to business as usual, it was recklessly spending more money than ever before.

I picked up the phone and called Newt's closest adviser, Joe Gaylord. I had great feelings for Joe since it was his campaign school that gave me the training I needed to get elected to Congress. Gaylord was crazy enough in 1993 to believe the Republicans could take over Congress and insightful enough to discourage me from ever hiring a campaign manager.

"You'd eat him alive." Gaylord laughed.

He was right about that, along with the notion that the Republicans could seize the majority. I gave him the news.

"Salmon just called. He's got seven votes."

Joe knew what that meant.

"Tell him to back off and support his Speaker."

"He's not backing off. It's over."

Joe let off steam for a minute, but he knew it all came down to simple math. Republicans had a six-vote majority. Matt Salmon had seven Republican votes against the Speaker.

The equation added up to the end of Newt Gingrich's political career.

Chapter Seven

The Easter Rising

NEWT GINGRICH MAY HAVE lost his job officially after the 1998 elections. But his fate was sealed months earlier when he lost support of the mob who had elevated him to power.

Washington is ruled by a mob mentality. So long as the mob senses invulnerability from its leader, the mob falls in line. This includes fellow politicians, lobbyists, D.C. social fixtures, and media bigwigs. But once a leader's wave of popularity crests and crashes, the Potomac Prince is rounded up and knocked from power with ruthless efficiency.

Think Paris, 1789. Or Washington, 1800. That was the

year John Adams learned that political power in American democracy has the consistency of vapor. Paraphrasing the modern philosopher Archie Bunker: *Washington power is like beer. It can be rented, but not bought.*

On a cold January morning in 1995, Republican congressmen rose rapturously to their feet while squealing with delight in tribute to incoming Speaker Newt Gingrich. The plump Georgia congressman had adopted a scorched-earth approach to Capitol Hill politics since first coming to Congress in 1978. The result being that by the time he took control of Congress, Newt Gingrich was the most feared and loathed figure in modern congressional history. But on January 4, 1995, Republicans throughout Congress paid tribute to the man who grabbed the House gavel for the Republicans for the first time since 1954.

Roars rose through cavernous congressional halls that reminded me of the chants I had heard on C-SPAN watching the British House of Commons. Other times that year I swore I was reliving a scene from *Planet of the Apes*. But that morning, the groans and howls reached a crescendo, and then melted into rapid-fire grunts of "Newt! Newt! Newt!" After five minutes or so, the Speaker-elect raised his hands while shaking his head, as if to say, "No . . . please . . . stop." But the cheers intensified. After ten more minutes of mad barking from the middle-aged Rotarian mob, Gingrich calmed the teeming masses and launched into a sixty-minute address on how we could help him save Western civilization.

Despite later objections to Newt's management style, Gingrich was as capable of holding his audience in rapt attention for as long as any public figure I have encountered. This includes Bill Clinton.

Unlike the baby-boomer president with whom Gingrich would be forced to share the political stage, Newt delivered sixty-minute romps through politics, pop culture, and new-age goop that focused more on wedge issues than self-serving parables. But unlike Bill Clinton, Gingrich never figured out how to boil big thoughts down into digestible sound bites. In modern political life, this flaw is usually fatal.

When our new Speaker did manage to spit out a memorable sound bite, the result was usually disastrous. Whether suggesting murdering moms were encouraged by Democratic agendas, or saying the government shutdown was by an Air Force One snub, Gingrich's verbal gaffes were quickly grabbed by an aggressive and hostile press corps and turned into front-page copy. Though he was already portrayed as the "Grinch who stole Christmas" even before being sworn in as Speaker, Newt's string of undisciplined remarks did little to help his cause in the press. In the end, Gingrich proved to be his own worst enemy.

When his first term as Speaker came to an end, House Republicans whispered to GOP staffers that they needed to keep their boss's mouth shut. But the advice went unheeded. Gingrich's poll numbers continued to plummet, and true to form, the pack of sycophants and parasites who made up the Newt mob slowly began distancing themselves from their former Republican messiah. Arianna Huffington, who has become the darling of the Hollywood left, was one of Newt's closest apostles while his star was still on the rise. During her ex-husband's California Senate run, Huffington told *Vanity Fair*'s Maureen Orth that she wanted to take the federal government in a "radical new direction" where private charities replaced public bureaucracies. Huffington was such a Gingrich groupie that in 1995 she set up a camera in the hallway outside my congressional office interviewing GOP congressmen on the virtues of Newt's Contract with America.

Like many around Gingrich, Huffington got the hell out of Dodge as soon as the bullets started flying. She headed west from Washington to Hollywood, where she somehow managed to convince Hollywood's most liberal stars that she was a left-wing populist. In the 2003 California gubernatorial recall election, she spent $100,000 mostly raised in Tinseltown before dropping out. As Maureen Orth noted in "The Importance of Being Famous," Arianna Huffington proves that among Hollywood stars, "there are new people to fool every minute."

But before Newt's betrayal, there were whispers. Unlike others who began talking behind the radioactive Speaker's back, Ways and Means chairman Bill Thomas was one of the first congressmen to blast Gingrich for his failures in running House operations efficiently. Talking in front of the Republican House Congress in 1997, Thomas took his former roommate head-on.

"Newt, you can tell us what is going to happen in America fifty years from now but you never seem to have any idea what we'll be doing on the House floor next week."

A dead silence hung over the hall.

Members' eyes began darting across the room, trying to decipher whether it was safe to offer a nod of approval. It was not. Old Bulls and committee chairmen sat stone-faced with their arms crossed, their body language suggesting contempt for a fellow chairman who dared to go against the party line. But Washington is a strange place, and these Old Bulls who had been saying much the same behind closed doors as Thomas said on the floor now frowned at his obvious observation. Later, when no one was around, they would slap him on the back for "saying what had to be said." One of the things uniform in politics is that speaking an uncomfortable truth is always discouraged by leaders of both parties.

While Bill Thomas's early critique was met with silence, publicly at least, the mob's was slowly moving away from Gingrich. Power abhors a vacuum, and many of Gingrich's lieutenants knew it was only a matter of time before last year's prince would become next year's exile.

Majority Leader Dick Armey was second in power behind Gingrich. He offered the Speaker nothing but effusive praise publicly, but by 1997 Armey began whispering to House "rebels" like Oklahoma's Tom Coburn and North Carolina's Sue Myrick that he was suffering from "Newt fatigue" and was "tired of cleaning up after him."

Sending such a signal to the already emotionally unhinged group filled by the likes of . . . well . . . me was like giving gunpowder and blasting caps to a Montana survivalist group. Something bad was bound to happen and it did.

Within a few weeks, South Carolina congressman Lindsey Graham called a "secret meeting" of rebels in one of Armey's conference rooms. Graham's office then alerted the Capitol Hill press corps to the time and place of said secret meeting, told them what secret issues would be discussed at the secret meeting, and gave them the secret count of members who had secretly agreed to attend Lindsey's secret meeting. Now that I look back on the debacle, it's a wonder Graham decided to run for the Senate a few years later instead of angling for a position in George Tenet's CIA. Predictably enough, a mob of reporters lined the halls leading into our secret meeting to depose of Newt Gingrich. It was yet another embarrassing moment in the long, strange saga of the Freshman Class of 1994.

It also added to the perception that we were the gang that couldn't shoot straight. Matt Salmon, representative of Arizona, and others knew that I would not join what was sure to be perceived as a confederacy of dunces in Dick Armey's office, but Salmon asked me to tag along as a personal favor since he considered himself to be in no position to bail out on his friends.

Going under protest, I tucked my head with the same shame I might bear were I walking into an N'Sync reunion concert. Once inside, I kept my arms and legs crossed, my head down, and my mouth shut. For two hours, members ranted, rambled, and, as always, got nothing accomplished. I dutifully waited for Salmon to make his final comments and then I left, passing the reporters outside without saying a word—a first.

Proving that once again no good deed goes unpunished in Washington, the next morning I found my face on the cover of *Roll Call* and was identified as the ringleader of this very public revolt. I immediately called reporters who wrote the story and demanded to know the basis for their reporting and why they didn't call me for confirmation. I didn't receive any good answers until a week later, when the same *Roll Call* reporter published a front-page puff piece on a member who had been the source of their false story.

It didn't matter in the end anyway since nothing was kept secret in Lindsey's secret meeting.

In a larger context, the episode proves that Washington

news outlets always protect their sources. Any newspaper editor or network executive who denies the existence of these "journalistic" payoffs is lying to your face. If you don't believe me, check past clippings on Secretary of State Colin Powell or Republican elder statesman James A. Baker III.

Like all others in the nation's capital, the Washington press corps protects their own until their own can no longer help them out. But the other side of this story is how those burned by media outlets save their best stories for other sources. After the offending story, I stopped giving *Roll Call* reporters any information regarding the brewing coup against Gingrich and instead started building a relationship with *The Hill*'s Sandy Hume. Young, brash, and conservative, Hume was dismissed as a "Nazi" by many of his young peers who covered Congress. Yet the only thing Hume ever did to be associated with the führer's party was write even-handed stories about a Republican Congress that was almost universally loathed by the press.

After returning to the Capitol from our July 4th break in 1997, congressional leaders began quietly approaching young rebels who had previously moved against Gingrich. Their message: The mob is ready to run Gingrich out of town.

When not meeting with us, Armey, Delay, and other House leaders were quietly plotting behind the scenes. Armey's meeting with Myrick and Coburn to discuss overthrowing Newt occurred on July 9, 1997. The next night, the House floor was electric as a small band of rebels quietly plotted their next move. Lindsey Graham ran up to me breathlessly and said, "It's happening tonight!"

"Why?"

"Delay said a newspaper's got the story and we better do it now or it won't happen."

Not one for meeting false deadlines, I turned to Matt Salmon.

"I don't want to look like an idiot over this Gingrich stuff again. What do you want to do?" I asked my friend.

We decided to get Delay alone on the floor and ask him.

"Tonight?" I asked Tom while walking beside him.

"Yep," he said quickly without looking our way.

An hour later, twenty-four House members met in Lindsey Graham's packed office to discuss the best way to remove Gingrich as our Speaker. We decided to make a motion to vacate the chair—a parliamentary move that boots the Speaker if he fails to receive a majority vote of members. Within Graham's office were all the votes needed to put Gingrich out of his misery.

By the time we moved against Newt, it was universally understood that for all his gifts, Gingrich was a drag on all Republican efforts. Regardless of what policy objectives we put forward, Bill Clinton, Democrats, and the press would slap them down by tying all House efforts to Newt Gingrich—who, by this time, had dismally low approval ratings. The writing was on the wall and it had less to do with Gingrich personally than the survival of the party. Even Karl Rove later declared to party insiders that George W. Bush would never have been elected president of the United States if Gingrich had not first been removed as Speaker. This political fact is beyond dispute.

But what remains disputed are the events of July 10, 1997. That night, as we malcontents sat in Lindsey Graham's cramped Longworth House office, Tom Delay told us that he would vote to vacate the chair and end Gingrich's Speakership.

At the same time, House Majority Leader Dick Armey and House leaders like Bill Paxon and John Boehner were nervously discussing what would come next. Armey said they had decided to have Judiciary Chairman Henry Hyde nominate him for Speaker. But for Armey, this joyous night was about to take a nasty turn that would forever taint his reputation and end his effectiveness as majority leader.

As Delay was wrapping up his business with the group of rebels in Lindsey's office, Oklahoma congressman Tom Coburn spoke up.

"We want Paxon to be the next Speaker, not Armey."

Whoops.

Coburn, who was an intelligent, well-meaning public servant while in the House, could also be a little stubborn when it came to adhering to the basic tenets of Politics 101. You don't divide your allies the night before your biggest political fight. Coburn did, and a few in Graham's inner office knew the coup could now unravel before it even began.

Tom Delay walked back to the room where Armey, Paxon, and Boehner were waiting. He gave Armey the bad news. He didn't have the rebels' support for Speaker. The Texas leader immediately went into action, warning Gingrich of the plot and making a speedy retreat to what Lisa Simpson once described as the last refuge of the truly desperate: religion.

Armey shocked Steve Largent the next morning by telling him that what we were doing to Newt was a sin and telling Largent how he and his wife had gotten down on their knees the night before and prayed for guidance. Jesus apparently provided grace aplenty because the next morning it was Newt and Dick against the world with Armey accepting Bill Paxon's resignation when the young New York congressman offered it to Gingrich.

For Armey it was a helluva conversion on the road to his political Damascus.

I have long suspected that when confronted with the unhappy news that he was once again going to be the bridesmaid instead of the bride, Armey probably muttered, "Jesus." Then he yelled "Jesus!" in disgust. And then with serenity that surpasses all understanding, Dick probably held a finger in the air, smiled, and said in a most prayerful tone, "Jesus."

While Armey was busy holding the Son of God in front of himself as a human shield, all hell was breaking loose on Capitol Hill. Word leaked out of the coup attempt as Gingrich and his lieutenants quickly moved to restore order. With Delay and other senior members implicated in the coup scrambling to save their political lives, the rebels from '94 huddled together to figure out their next move.

Most were disheartened and did little other than lick their wounds and head back to their home districts. Matt Salmon and I encouraged the group to fight back and tell their stories to the press. Largent wasn't in the mood to talk to anyone, while Coburn and others saw no upside in a public relations counteroffensive.

"They won't listen to us now," was the most popular refrain.

But Salmon and I disagreed. That Sunday we went on ABC's *This Week* as I worked quietly behind the scenes using other means to get the real story out.

Early the next week, Sandy Hume broke the political story of the year when he gave readers of *The Hill* a minute-by-minute rundown of what happened during the days leading up to the coup. The Hume story shocked congressmen and senators, and destroyed Dick Armey's reputation among his peers by painting him as a shifty operator who used religion as a political crutch in times of trouble.

Armey was livid at *The Hill* story and used the next Republican caucus meeting to dismiss the Hume article as a "pack of lies." The denial was ridiculous since, by now, Armey's actions that night were a well-known secret inside the caucus. Sandy Hume's story simply made Armey's mendacious actions known around Washington and the world. Still, Armey's actions were too much for an embittered Lindsey Graham to take. In the middle of the caucus, Lindsey started throwing chairs out of his way to get to the microphone while screaming that Armey was a liar. Fortunately for the majority leader, Graham was physically restrained before reaching the podium. But Florida congressman Tillie Fowler comforted Graham.

"Don't worry, Lindsey, we all know the truth."

That was precisely the problem for Dick Armey, but fellow Texan Tom Delay was facing his own set of challenges. Unlike the other House leaders, Delay had actually met with our group. He was in the room with twenty-four rebels plotting the overturn of Speaker Gingrich. Now Gingrich's lieutenants were calling for "the Hammer's" head.

But Delay chose that time to take an approach considered radical by many on Capitol Hill. Instead of following Armey's denial strategy, Tom Delay decided to tell the truth. Standing before the Republican conference, Delay tearfully confessed his sins and threw himself on the mercy of the Republican caucus. And while most members felt charitable toward Delay, Newt Gingrich did not.

Soon after Delay's confession, Rules Chairman David Drier approached me with a message from Newt meant for the rest of the rebels as well.

"Hey, Joe," Drier began, "you're a reasonable guy, and that's why I came to you to explain that someone's going to have to pay for what went on this past week." Drier paused. "And that someone is Tom Delay. He's got to go."

I smiled back at David, whom I had considered a good friend since coming to Congress, and said, "That's fine, David. But tell Newt that if he dumps Delay, there will be a motion to vacate the chair within five minutes. And this time we'll have the votes to finish him off for good."

Drier flashed a quick smile, knowing we had the numbers on our side, and said, "Thanks, buddy. I'll pass it on to Newt."

He did and Delay survived—but because he told the truth. And this was one of those wonderful times that telling the truth in Washington worked to a politician's advantage. Delay's power base within the Republican Party and on Capitol Hill would only continue to grow after the failed coup attempt.

For other parties involved there would not be a happy ending. Dick Armey sent a letter to his fellow congressmen ten days after the coup attempting to salvage his reputation. Armey wrote, "I just want my good name back," and defiantly declaring, "I'll be damned if I'll let my name and honor be destroyed." But after eleven days of blasting Sandy

Hume's article in *The Hill* as a pack of lies, the majority leader backed down—sort of.

"When I went before the Conference about the article in *The Hill,* I spoke what I believed to be the truth. . . . The reaction by Lindsey Graham frankly surprised me, and I began digging to find out why he and others would call me a liar."

Armey went on to write that he understood why we were so upset but again denied any involvement in the botched coup. His unpersuasive written denial would seal the majority leader's fate. The conference would forever consider him a hypocrite and a backstabber who didn't even possess the decency to admit he was a hypocrite and a backstabber after being caught red-handed engaging in both Capitol Hill sports. With his reputation beyond rehabilitation, Armey and crew began looking for opportunities to exact revenge on Sandy Hume. Unfortunately, they didn't have to wait long.

Following the *Hill* piece, Hume complained that Armey and his staff were doing whatever they could to damage Hume's career. But Sandy had no need to worry. Besides having one of the most respected figures in Washington journalism as his father, Sandy Hume had also just broken a political story of gigantic proportions. Offers from top newspapers, television, and magazines started pouring in. But it was not to be.

Sandy passed away on February 22, 1998, at the age of twenty-eight. His tragic and untimely death shocked his friends and family members, especially as Hume's star was on the rise and the entire world was about to open up before him. Unfortunately, Hume's death presented an opportunity for revenge to some in the Armey camp.

Soon after Sandy's death, I received a call in my congressional office from a *Wall Street Journal* reporter. She whispered a question I knew she was ashamed to ask.

"Congressman, I'm sorry about Sandy. I know you two were friends . . . but . . . um, I have to ask you this. Someone in Dick Armey's office just told me Sandy Hume got the coup story by having a gay affair with Bill Paxon. Both knew the story was coming out and that's why Paxon resigned and why Sandy killed himself."

Bill Paxon had unexpectedly resigned from politics a day after Sandy's death.

I sat at my desk, rage growing by the second toward this reporter, Dick Armey, and the sleazy business of politics. For the first time since the coup began, I measured my words and held my anger.

"I don't know who Sandy slept with and I don't care. But I do know Sandy didn't get his story by having sex with Bill Paxon. Sandy got his story by eating ribs in Shirlington with my ten-year-old son and me. I told Sandy everything about the coup from beginning to end to show everybody what really happened. Bill Paxon didn't give Sandy the story. I did."

There was silence on the other side of the line. Finally, the reporter apologized for bothering me and thanked me for my time. She and most other Washington reporters ignored the story that spread around the Capitol like a virus.

Dee Dee Myers would tell me at Sandy's funeral that the slur against Hume was the single most disgusting thing she had ever encountered in politics. Those around us nodded in agreement that while winning in Washington was always considered a blood sport, the slur on Sandy Hume was more like a political terrorist attack. And in this tragedy the collateral damage was not to the reputation of the intended target. Everyone knew Armey's staff was lying. Instead, it was Sandy's family and friends who had to endure the unspeakable savagery of the attack.

I wonder, "What would Jesus do?" and if anyone in Armey's office even asked.

Chapter Eight

Fat White Pink Boys

PRESIDENTS, SENATORS, and congressmen do not run Washington.

Instead, the American government is controlled day in and day out by the same Fat White Pink Boys I spoke about earlier. They populate the ranks of all congressional staff, D.C. lobbying firms, bureaucratic agencies, congressional committees, political consulting firms, White House cubicles, and just about every other desk at national party headquarters.

Fat White Pink Boys, as you might have guessed, are usually male. They gather in herds at Capitol Hill watering holes wearing suspenders, starched white shirts, and Nixonian flop sweat. Male pattern baldness is an identifying feature of many Fat White Pink Boys.

I have studied this confounding species for years and have come to realize that their goal is not to live a life committed to public service—or even to their Washington bosses. Instead, the goal of any and every Fat White Pink Boy is little more than keeping his bureaucratic job and receiving his next paycheck from the Federal Treasury. These two objectives require the "Pinkies" to put their personal agenda ahead of the elected representative they presumably serve, so it goes without saying that balanced budgets, tax cuts, and government downsizing are frowned upon by these guys. After all, the more money, power, and influence that remains in Washington, the more likely they will get next month's paycheck.

When I served in Congress, I was initially shocked to find staff members were willing and eager to go behind my back to try to undercut certain legislative goals. When caught red-handed, these Fat White Pink Boys would point fingers furiously at staff members or other members of their own species. And as you know from your own experiences at work and at school, Fat White Pink Boys will cut the throat of one of their own if that is what is required to survive in the workplace. My experience in the media has likewise shown me that "Pinkies" populate TV networks and print newsrooms in equal abundance. The names may change but the suspenders all look the same.

I always suspected Fat White Pink Boys held secret meetings after the human population of Washington went to bed, though I must admit I never unearthed solid evidence to confirm my suspicions. Wisely I did take copious notes while observing these creatures and can undeniably tell you that Fat

White Pink Boys live by a secret code that's never been seen by human eyes until now.

The Secret Code of the Fat White Pink Boys

1. Don't let convictions stand in the way of power.
2. Never pass up the chance to kick someone when he's down.
3. Never root for the underdog.
4. When in doubt, lie.
5. No ass is ever too foul to kiss.
6. Never show mercy to the weak.
7. If they can't help you, don't return their call.
8. Power trumps family every time.
9. Power trumps friendship every time.
10. Shamelessness: the most treasured virtue.
11. Stand by your man—until betrayal makes more sense.
12. Beware the man with convictions and backbone.

My first encounter with the Fat White Pink Boys occurred years before I was sworn into Congress. It was during my quixotic campaign to abolish the student government association at the University of Alabama that I came into contact with these cretins. While I can now admit that I saw that political campaign as a chance to verbally abuse the most pompous and pampered frat boys Birmingham and Montgomery had ever produced, there was actually a larger issue at play. The

University of Alabama student government was run by a secret organization organized by Jefferson Davis in the nineteenth century. The group called itself "The Machine"—and for good reason. They controlled every aspect of campus life and ruthlessly protected their political turf. Independent-minded individuals like myself who dared cross the Machine soon found their phones tapped, their cars run off the road, or their heads beaten in.

The university administration rarely took issue with the group because the family members of the Machine were some of the school's largest contributors. The fraternities and sororities protected this exclusionary and racist system because election to this student government was the best predictor of future success in Alabama state politics. It was a brutally efficient way to preserve Alabama's political caste system and would probably still be going strong until this day but for the fact that one night some goons who were card-carrying members of the Machine beat up the wrong girl.

Minda Riley was a sorority sister who decided she wanted to follow in her older brother's footsteps and run for president of UA's Student Government Association. But Minda wasn't the person the Machine wanted to run that year. Since she belonged to a sorority that was tied into the Machine, few in the secret organization took kindly to Minda's stubborn belief that she should be SGA president. So rather than try to talk sense into this strong-willed girl, a thug broke into her apartment, beat her, and left this Phi Mu sorority member with a warning: "Get out of the race or else."

I always considered myself a contrarian—though others simply passed me off as a pain in the neck—so I respected Riley for standing her ground before and after the beating. She went to the police and to the press, and soon university officials had little choice but to disband the Machine and the Student Government Association that had bred such contemptible behavior. Ten years later, her father, Bob Riley, joined me in Congress, and then in 2002 was elected governor of Alabama.

Minda Riley's experience didn't surprise me because I had crossed the powerful Fat White Pink Boys masquerading as college students a few years earlier. While other independents preached reform to no avail, I decided to skip the whole good-government bit and go straight for the Machine's jugular. I pledged to abolish the secret society and their wholly owned subsidiary, the University of Alabama's SGA.

I took a less than nuanced approach to this personal crusade. I slammed these spoiled Fat White Pink Boys as country club elitists who propped up an exclusionary organization that systematically left blacks, foreign students, and independents not pledged to a sorority or fraternity unrepresented on campus. When I chose to run for SGA president in 1985, white students had won every campus election in the preceding five years.

The outcry was fierce.

Many professors chided me during class for running a divisive campaign. Administration officials gravely warned that my student protest efforts would destroy any chance I might have of getting involved later in life in state politics. I gave them the same sneering laugh that I used on the Pinkies hired by Clinton or Gingrich. Meanwhile, my left-wing professors who survived the '60s finally found something they agreed with me on.

"Dude," one literature professor asked, "why are you telling the truth about these clowns?"

"Because what they are doing is bullshit," I replied, craftily adopting his native tongue.

"Cool, man," my most excellent professor responded with a smile that told me I would be making A's in at least one class for the remainder of the semester.

Others who ran against the Machine faced bodily harm over the years, but my only brush with violence came late one night during the campaign. It was at this time that one of the hottest women on campus (sadly a member of a Machine sorority) chose to take note of what a fine, young specimen I was. Following an ambush strategy that some fraternity bonehead must have plucked from the script of *Animal House,* this piping-hot sorority girl dialed yours truly and began whispering into the phone.

"Joe," she moaned from deep within what I have long suspected was a massive, heaving chest, "I feel so bad about how those boys have been treating you. Can we get tahgetha' tonight so I can tell you face-to-face how much I respect you for your courage."

"Where?" I asked, still sleepy but able to focus on what I knew just had to be her massive, heaving chest.

"How about behind the stadium, across the street from the Alpha Chi house?" she whispered.

Fortunately, I awoke from a hormonal-induced fog long enough to imagine the sounds of billy clubs and brass knuckles slamming into the palms of angry DKEs and SAEs. I knew at this point in my college life that it was not simply bad luck that had prevented rich, beautiful sorority girls who looked as if they stepped off the pages of those cursed Ralph Lauren catalogs from waking me up with phone calls begging for late-night sex.

"I think I'll pass tonight. How 'bout giving me a call after the election?"

The other side of the phone line went dead. Much like I suspected of President Bush's budget director Mitch Daniels, I had a pretty good hunch that Miss Perfect would not be calling me back again anytime soon. But this incident and the entire campaign was a perfect introductory course preparing me for the ways of the Fat White Pink Boys I would encounter on Capitol Hill.

My first brush with the Fat White Pink Boys in Washington came at the end of a long week leading into the 1998 Congressional Easter Recess. For months our fight to hold down spending on appropriations bills had caused us to skirmish with Bill Clinton, Newt Gingrich, and many of our own Republican brethren.

Many of my closest friends were damn near crazy by Washington standards, but there was nothing half-crazed about my approach to reform battles. I had drained the Kool-Aid bowl dry. I found myself along with the small group of mavericks with whom I had aligned once again locked up in a tight-quartered leadership office. Dick Armey was lecturing us on the subject of playing well with others. It was Armey's job to play the role of the bad cop trying to reason with the irascible bomb throwers.

Armey told us in a mournful tone that while he hated spending more money on appropriations bills, we couldn't do a darn thing with that liberal Bill Clinton. It was a familiar refrain, but since this spending fight was over money Congress would give itself to run legislative branch operations, the pitch didn't sell. The Legislative Branch Appropriations Bill is made up of requests directed at managing Congress, running committees, and keeping the House and Senate in working order. Suggesting the president was responsible for this bloated congressional bill was a farce.

"Nice try, Dick," Arizona congressman Matt Salmon said with a laugh, "but Clinton ain't gonna veto our own budget."

Mark Sanford, John Shadegg and I flashed Armey a knowing grin and sweat immediately started glistening from the foreheads of the Fat White Pink Boys lining the walls of Armey's office.

Armey grumbled something about "crazy bastards" and was quickly ushered out by the pack of Pinkies who were sure to tell the majority leader what an inspirational message he had delivered to those deranged anarchists.

Three days later, on the eve of the long Easter recess, Gingrich called for a vote on the issue. Just as I did, Salmon, Lindsey Graham, Steve Largent, Mark Sanford, Steve Chabot, Tom Coburn, and five other Republicans held firm by voting down the bill. By killing Gingrich's funding request for his own House account, we had effectively shut down the U.S. Congress.

Gingrich was outraged.

The House Speaker immediately sent his pack of Fat White Pink Boys out to sweep the House floor, the cloakrooms, and every Capitol Hill bar to deliver each and every Republican member into an emergency caucus meeting. The agenda for that evening was the hanging of the eleven House mavericks.

As our group slowly wound our way through the bowels

of the Capitol complex toward a grim basement meeting room called HC-5, we found ourselves walking through a gauntlet of Fat White Pink Boys who worked for Gingrich, Armey, Delay, and various committee chairmen. These Hill staffers quickly picked up on the scent of blood, sweat, and fear coming from our group and started shouting at us. One staffer who had spent the previous three years kissing up to me stepped out from the line and grabbed my arm saying, "Nice job, Scarborough. You just made me miss my vacation."

I turned to find Salmon for reassurance, but he had been accosted by another pack of Pinkies and dragged into their huddle. By the time we finally reached the room, Gingrich was on his feet yelling for an aide to call roll and to have the sergeant-at-arms drag in any members not yet in the caucus room.

"I think we burned one bridge too far," I whispered to Salmon.

But the normally unshakable Salmon wasn't breathing. He and the other ten dissidents were frozen in their chairs awaiting their slow and gruesome death.

After the usual sycophants, apparatchiks, and party hacks made their opening remarks, Gingrich stood up and shot us a glare that would've made John Wayne flinch. For the next thirty minutes the Speaker blasted away at the Gang of Eleven for being stupid, shortsighted, and self-absorbed. For once, I wasn't slouched in my chair reading *Roll Call*. For the first time since being elected to Congress, I was actually scared witless. I believed I had finally crossed the line separating political courage from rank stupidity.

The Fat White Pink Boys slowly worked themselves into a

rich lather, as Gingrich continued wailing away at our terrified group. They started hissing at us like hyenas while the 224 other Republican congressmen cast icy glares in our direction. Then Gingrich made a fateful move that began the eighteen-month march to his political Waterloo you have already read about. He voluntarily yielded the floor to the rebels.

"I'd like these self-righteous members—who think they're so smart—to get up here and explain why they just gave Dick Gephardt and our enemies this victory."

If Gingrich made his decision to toss the microphone our way believing we'd be too frozen by fear to leave our seats, his calculation was a wise one. His blistering speech, the Fat White Pink Boys' gauntlet, and the open contempt our fellow Republicans showed us had subdued even the cockiest maverick from the once-radical Class of 1994.

But Gingrich didn't count on one thing. He didn't count on Steve Largent.

Largent, an NFL Hall of Fame receiver, was the golden boy of our group. Steve was universally respected. He was loved by his peers and had even then recently been named one of *People* magazine's most beautiful people—a moniker, trust me, that would never be placed on my misshapen head.

Golden Boy Largent rose from his chair and slowly walked up to Gingrich, taking the microphone from his hand while ten of us terrified "bad boys" of Congress sat subdued in our seats. A look of surprise briefly crossed the Speaker's face, but he yielded the floor and moved quickly to his corner.

"Newt," Largent began, "when I signed my first deal with the NFL, I shook the owner's hand and promised to do my best to make him proud. A few years later, NFL players went on strike, but I was one of the only members of my team to cross the picket line—knowing I was enraging two-hundred-and-fifty-pound linebackers who could—and now would—try to kill me the next time I caught a pass over the middle.

"Twelve years later you asked me to travel to Washington, stand on the steps of the Capitol, and sign the Contract with America—promising my voters that I would cut congressional staffing and slash our budget. When I did that, I looked voters in the eye, shook their hands, and promised them I would make good on my promise.

"And like the promise I made to the Seahawks' owner," Largent said in a low, steady voice, "I made a vow to the people in my district. And I intend to keep my word."

Largent moved his eyes from the formerly enraged Republican caucus, who now sat in rapt attention as he turned to face Gingrich.

"So, Newt, if I wasn't intimidated by a two-hundred-and-fifty-pound NFL linebacker, I'm not going to be intimidated by you. None of us are," Largent said, pointing toward our group, who just got airlifted out of the Valley of Death by the golden-boy congressman at that moment. I swear I saw a dove descend from the heavens and land on Largent's shoulder.

Gingrich was now slumped in his chair while the eleven of us who followed our campaign promises regained our composure, our cockiness, and our swagger long enough to survive the most frightening hour of our collective political career.

Lindsey Graham took the microphone next and got everyone in the basement room laughing with the first line out of his mouth: "I just want it to be known for the record that, unlike Steve, I can be intimidated."

Tom Coburn followed with a very soft-spoken and humble talk about why he was moved to join our revolt. While he spoke, Appropriations Cardinal Hal Rogers (R-Ken.) stood five feet from Coburn and yelled, "You're a self-righteous hypocrite!" Rogers continued yelling insults and profanities during Coburn's speech, but the Oklahoma doctor kept his cool. The outburst by this Old Bull only hurt Gingrich's cause more. This fight was over. We had won.

Steve Largent had saved our political hides while unwittingly laying the seeds for Gingrich's political destruction in just under five minutes.

The meeting broke up soon after, and the same Fat White Pink Boys who earlier whispered "Dead man walking" as we passed through their gauntlet, now slapped us on the back.

"You got balls, Congressman. Great job," one said, as I turned around to see the same Fat White Pink Boy who had cursed at me an hour earlier for making him miss his flight. Now he was smiling and offering his hand in admiration.

Chapter Nine

Why Washington Always Wins

SOME STORIES JUST don't have happy endings.

But we Americans suffer from an optimism others interpret as naïveté. Maybe that's why we sit in front of our TV sets gazing mindlessly at horrific video images looped by cable news outlets. Maybe we believe if we watch the space shuttle replay enough times it will race into the heavens instead of violently crashing to earth. Or maybe, just maybe, the second plane will miss the South Tower on the one hundredth viewing of that grim video clip of September 11th.

Soon after John Lennon was murdered, a writer for *Rolling Stone* told of his struggle to come to terms with the former

Beatle's death. He followed news reports about the slain rock star with a religious zeal. Soon enough, though, he reminded himself that this story would never have a happy ending. Like *Ben Hur,* the movie would always end the same. The bad guys would win and the hero would be vanquished.

When it comes to cinematic political fare, Americans have had a long love affair with *Mr. Smith Goes to Washington.* In that 1937 classic an idealistic Jefferson Smith is appointed to the United States Senate and barnstorms Washington, hoping to make a difference. But the young senator is soon confronted by the awesome might of Washington's political machine. When Smith—played by Jimmy Stewart—stands up to the party bosses, they unleash the political attack dogs, hoping to destroy the reputation of the young reformer. But our Mr. Smith fights back, defeats the political bigwigs, and watches his leaders confess their errors. He even wins the girl.

It is a nice Hollywood tale, but unlike Mr. Smith or Perry Mason, the guilty in Washington rarely confess. In 1994, dozens of political innocents in my freshman class met hardened congressional insiders on both sides of the political aisle who had no intention of reforming their gluttonous ways. Unlike Jimmy Stewart's Hollywood struggle, the dramatic stand by the Class of 1994 against the status quo never took place on the floor of Congress when the cameras were rolling. There were no makeup artists at the ready to drench our eyebrows between takes with dramatic flop sweat. Instead, the showdown between political mavericks and the Washington establishment was far less cinematic. By the end of our movie, Washington power brokers won the day. Sadly, in real life they almost always do.

The Reagan Revolution changed the terms of the economic debate during the 1980s, but as former Reagan budget director David Stockman outlined in *The Triumph of Politics,* political reform movements are usually doomed to fail from the start. In Reagan's Washington, that meant Congress went along with the White House's tax cuts but didn't have the guts to curb runaway spending increases. Ironically, it was Congressman Dick Cheney who signaled the end of the Reagan Revolution in 1981, when he told administration officials that Congress lacked the will to cut spending at the rate proposed by the new Republican president. Then, as now, Congress passed the president's tax cuts but ignored all calls to rein in federal spending. Too many Republican congressmen running away from tough economic choices these days are the same storm troopers who once pledged to reform the Imperial Congress.

So why did these Visigoths become just one more class of custodians determined to preserve Capitol Hill's status quo? And why has government spending exploded at record rates under a self-described conservative Congress and Republican president? The second question is easier to answer.

My experience in Washington has taught me that one-party rule can be a dangerous thing. The system of checks and balances installed by our founding fathers atrophies when a powerful president and his Capitol Hill allies have no political incentive to hold each other accountable. The first two years Bill Clinton served as president alongside a liberal Democratic Congress proved to be a political disaster for Democrats as well as the country. Neither Clinton nor his Democrat partners in Congress felt compelled to moderate the most extreme elements of the party agenda. Within two years Clinton Democrats passed one of the largest tax increases ever and tried to socialize one-seventh of America's economy. Even after running as a new Democrat in 1992, ruling like an old one in 1993, getting trounced in his first midterm election in 1994, being forced to sign Republican Welfare Reform Bills in 1996 and balanced-budget agreements in 1997, Bill Clinton took credit for the success of those Republican policies in his 2004 memoirs while suggesting his historic tax hikes and socialized medical schemes were great ideas derailed by right-wing extremists running the National Rifle Association and the Christian Coalition.

Is Bill Clinton still crazy after all these years? No. He is just a partisan Democrat who succeeded because our Republican Congress checked his worst instincts. As with Clinton and the Democratic Congress in 1993 and 1994, these days George Bush and the Republican Congress rarely cross each other on issues both foreign and domestic for fear of hurting "the bigger cause."

Here's a little Washington secret: The bigger cause among party leaders usually means re-electing your president and electing even more members from your own party to Congress.

Though I have supported George W. Bush's presidential efforts since 1998, I agree with Washington observers who believe the national deficit would be smaller if a Democrat had been elected president in 2000. This is certainly not because Democrats like Al Gore are more conservative on economic matters than George Bush. They are far from it, actually. Their multitrillion-dollar plan to socialize medicine in 1993 is a good example of that. But just as only an anti-Communist like Richard Nixon could make peace with China, only a conservative like George W. Bush could pass Ted Kennedy's education bill through a Republican Congress.

Republican leaders on the Hill would never let Al Gore pass the Kennedy education monstrosity, or the largest corporate welfare bill for agricultural conglomerations, or the economically reckless Omnibus Appropriations Bill of 2003. These big-ticket items would have been dead on arrival in this Republican Congress but for the fact that the Republican president wanted them passed. By now it is obvious that the current crop of Republican leaders in Washington are incapable of curbing their appetites for your tax dollars. There is no system of checks and balances protecting American taxpayers.

One depressing example of this dynamic at work came while the World Trade Center's ashes were still hot. Soon after September 11th, Congress passed a $318 billion defense bill in response to the national challenges created by the terrorist attacks. Nearly $10 billion was tacked on to the September 11th bill by members of Congress hoping to bring a few extra bucks back to their home states. According to Citizens Against Government Waste, members did this by tacking on "riders" to this critical bill and directing that money back home.

The mere thought of politicians using the deaths of three thousand Americans as an opportunity to pass another piece of pork through Congress may seem unimaginable to some, but to those familiar with the ways of Washington this seemingly shocking spectacle was nothing more than business as usual. A veteran congressional aide who called himself "Spartacus" posted a scathing indictment of such behavior on the Internet and invoked the legacy of Mr. Smith. His conclusion was depressingly similar to mine.

"Mr. Smith is dead, and no one in Congress, not even the very few who pose as reformers, has any regrets."

While I admit that some of my congressional friends still serving in Congress have lost their stomach for reform, the system that broke their will is a daunting minefield to navigate. And few survive with their backbone in check.

Republicrats

The CATO Institute recently released a paper suggesting that when it comes to government spending, the two political parties are indistinguishable. The American people are left with one "Republicrat" party—fueled by free spending and undisciplined ways.

The evidence supporting this cynical conclusion is compelling. Take for instance the research that measured Real Discretionary Outlays over the last forty years. Real discretionary outlays are the amount of money Congress chooses to spend but does not include programs like Medicare and Social Security, which grow naturally as more and more of the population ages. CATO disclosed that the biggest jumps in discretionary spending over the last forty years occurred under both President Lyndon Johnson, architect of the Great Society, and President George W. Bush.

The education bureaucracy of Washington is one of the fastest-growing domestic agencies fueling D.C.'s money machine, and much of that explosive growth can be traced back to the No Child Left Behind legislation of George W. Bush and

Ted Kennedy. The bill itself can be traced to the president's failed charm offensive launched in 2001 to win the heart and mind of Senator Kennedy. The liberal icon was invited to the White House to watch movies, discuss legislation, and find ways to work with the Bush administration. The result of this odd coupling was the education package that many Republican congressmen quietly called "Leave No Bureaucrat Behind."

While this strange political paring cost taxpayers billions of dollars, in the end the president received very little support in return. Senator Kennedy spent the succeeding years accusing the president of sending young Americans to die in Iraq as a way to win votes. Mr. Bush's new friend also charged that Operation Iraqi Freedom was nothing more than a Bush political stunt hatched in Texas. But Kennedy's insults did not stop with the commander-in-chief. In April 2004, he took to the Senate floor to compare American troops to the Iraqi thugs who slaughtered hundreds of thousands of Iraqis under Saddam Hussein's rule.

Some charm offensive, huh?

This political slow dance with Ted Kennedy was only one chapter in the story of how Washington ran up record deficits. In his 2004 State of the Union address, President Bush also asked for $300 million to help ex-convicts find work, $120 million for teacher training, and $250 million for job training. And he asked for twice as much money to fund government-managed abstinence programs. It's clear that there is a philosophy governing much of this spending. And as I explained earlier, President Bush is using the federal government to promote conservative causes.

Don't misunderstand me. I agree with most of the president's goals, but funding a conservative agenda through federal bureaucracies is an agenda that is incompatible with the values of Jefferson, Reagan, and the Class of 1994. When the government grows, America's freedoms decrease. When taxpayer money is taken out of Florida to fund an education bureaucracy in Washington, D.C., that is tax money that will not go directly to classrooms, teacher salaries, or local school boards. Maybe that's why Ronald Reagan used Thomas Jefferson's doctrine that the government that governs least governs best as his abiding philosophy.

If politicians want to increase spending to promote abstinence among teens, then it is only reasonable to assume they will cut spending somewhere else. This is not rocket science. In fact, it's not even University of Alabama math. Whether you are in Tuscaloosa, Alabama, Pensacola, Florida, or Washington, D.C., one plus one still equals two. Taxpayer money is a limited and precious resource, and even if the White House wants to use that resource to further the most noble goals of the administration, it should be forced to use less of it on something else.

Taxpayers' dollars are the most powerful tools a politician has. That is why congressmen on both sides of the aisle promote their agendas by throwing trillions of dollars around every year. Every time the government cuts a check it may not be to fund a pork barrel boondoggle or to promote a corrupt scheme that benefits a special-interest group, but you can take it to the bank that secret horse trading and backroom deals that have very little to do with America's common good drive more than a few of these political arrangements.

But Republicans cheering wildly for their team should not

fool themselves into believing their agenda is consistent with the values of the Republican Party of Ronald Reagan. It is not. In fact, this most recent fleecing of America is a betrayal of everything the Republican Party has been saying about fiscal discipline since Reagan burst onto the national stage with his memorable 1964 endorsement of Barry Goldwater.

The Leasing of America

This is not to say that all Republicans have joined their fellow Democrats by turning their backs on American taxpayers. Arizona senator John McCain continues to hold his fellow members accountable for their wasteful ways and has led several recent investigations concerning Washington waste, fraud, and abuse. After spending two decades on Capitol Hill, Mr. McCain will tell you that if congressmen ran their businesses like they run your government, they would be bankrupt in a week.

Imagine your reaction if a used-car salesman offered you this sweet deal: You can lease a car for $15,000 a year for the next ten years. Then at the end of the ten years you'll have to pay another $10,000 to fix up the car, make it look brand new, and give it back to the dealer. Unless you have employed a former Enron accountant, you wouldn't walk away from that deal, you would *run*. But when Congress was faced with a similar proposition that involved billions of dollars instead of just thousands, they jumped for joy over the offer.

The deal in question didn't involve cars, but rather airplanes. Big, expensive airplanes. Boeing pushed a deal to lease

767 airplanes to the Pentagon. These planes would be used as refueling tankers and would be transferred in a lease paid by American taxpayers for ten years. Each plane cost $150 million, but at the end of the ten years, the Air Force would also have to spend an additional $30 million per plane to reconfigure each for commercial use before giving it back to Boeing Corporation.

The Senate passed this extraordinarily stupid leasing deal by a vote of 94 to 2. The House rolled over on a 408-to-6 count. But the outline for this bizarre plan begs the question, What, aside from sleep deprivation, could explain this lapse of judgment?

I suppose we could blame it on lack of sleep, or perhaps maybe we should focus instead on the $4.5 million Boeing and its executives gave members of Congress over the last five years. The Boeing scam was part of the 2002 Defense Appropriations Bill that proves more clearly than almost any other military bill before it that national defense is a safe refuge for billions of dollars in corporate-welfare handouts.

September 11th was obviously an economic blow to the airline industry, and, consequently, they received their own bailout package, courtesy of your tax dollars. But companies like Boeing, which manufacture airplanes, were also hurt. Writing in *U.S. News & World Report*, Julian Barnes described how Boeing announced massive layoffs on September 18, 2001. And how within weeks, the chief lobbyist for the company had met with Alaska senator Ted Stevens, who, at the time, was the ranking Republican on the powerful Appropriations Committee.

Soon the Boeing leasing plan was born. And when the bill

hit the Senate floor, Stevens went to work behind the scenes. Barnes reports:

> Rick Santorum, the junior senator from Pennsylvania, turned to Stevens, his fellow Republican. "Why couldn't the Air Force keep the 767s," he asked his older colleague, "after the lease was up?" Stevens, wearing his favorite Incredible Hulk tie, shook his head. "We can't do that," he said. "It will queer the deal." Santorum dropped the question. Stevens moved on. Later that day appropriators slid five Santorum amendments into the appropriations bill. The amendments added $18 million in defense projects. All were earmarked for Pennsylvania.

Why was Stevens, a senator from Alaska, so eager to cut deals on behalf of Boeing? Did he truly believe the leasing deal was the best way to get the Air Force the refueling tankers it needed? Perhaps, or maybe he was inspired by the 2002 fundraiser Boeing hosted for him in Seattle. Boeing executives ponied up $22,000 to his campaign for re-election, making the powerful Senate chairman just one of the many senators and congressmen whose campaign accounts were stuffed with money by the military-industrial conglomerate.

I spent a great deal of time attacking the Boeing deal on MSNBC, and Senator John McCain led the charge on Capitol Hill to kill this criminal conspiracy before taxpayers got stuck with the $10 billion bill. Congress eventually bowed to public pressure and slashed $5 billion from the package. But sadly, there are hundreds of smaller scams resembling the Boeing deal that are scattered throughout the budget process that have helped rack up record deficits and a $7.4 trillion debt.

For instance, there's the tax credit for "synthetic fuel" that costs the taxpayers between $1 billion and $3 billion a year. It's supposed to fund the development of a synthetic-fuel alternative to foreign oil. In practice, big energy companies build production facilities that slightly alter the chemical composition of coal. Then they write off these facilities as "synthetic fuel" research. Last summer, the IRS tried to end the abuse, but the companies that benefit rallied to save the tax break.

A group of these companies hooked into the synthetic-fuel industry hired a lobbying firm called Clark Consulting. The firm employs former Joint Committee on Taxation chief of staff Kenneth Kies, and former Ways and Means staffers Jayne Fitzgerald and Tim Hanford. Another group of interested parties hired the law firm of Hunton & Williams, whose rolls are similarly stacked with former Hill staffers and members of Congress. The goal was obviously to keep the corporate sweetheart deal on the books. And soon enough, the IRS backed down.

I wonder if a lobbyist could stop you from having to pay taxes?

I certainly don't want to sound like a speechwriter for Ralph Nader, but the fact is, both the Democratic and Republican Parties are equally beholden to corporate America. And they find all kinds of ways to disguise what they are doing. Under President Clinton, $1.5 billion went to the automobile industry for fuel-cell research despite the fact there is little hope that anyone will be scooting around your hometown in a car run by fuel cells anytime soon. But let us assume that one of those fuel-cell cars makes it to a car lot before my

Boston Red Sox win the World Series. Will the Big Three car corporations in Detroit be kicking back the government's share of the profit?

Don't hold your breath.

Still, President Bush wants to give another $1.7 billion to the automobile industry for more taxpayer-funded corporate research to unlock the secrets of these vexing little fuel cells. If this corporate-welfare plan passes, it'll be for the same reasons the Boeing deal went through: big business using money and connections to exert big influence on politicians.

Of course, appeasing corporate interests is not the only motivator behind these giveaways. Money for corporations is, indirectly, also money for someone's home district. As the *Weekly Standard*'s Fred Barnes pointed out in his account of the Boeing deal, the planes in question were to be manufactured in Washington State, fitted as tankers in Kansas, and possibly based in North Dakota. So guess which senators were the most critical in passing this budget-busting bill?

Good guess.

Connecting the Dots

Every member of Congress wants to bring home money to their state, and there's nothing inherently wrong with that. Federal money leads to local jobs, which results in happy voters. Unfortunately, each member of Congress faces the same problem—each representative only gets one vote. Consequently, in order to get anything done for their district, they need to beg and barter.

There's no better example of how these backroom bargains make the legislative process spiral out of control than the farm-subsidy system. Since the Great Depression, agriculture has been subject to a complex web of centralized regulations, price controls, quotas, and subsidies. The cost to American taxpayers in lost efficiency and direct giveaways totals billions of dollars. It is a socialized agricultural system in certain regards equal to Joe Stalin's five-year plans—minus the forced starvation and slaughter of 30 million or so peasants.

In 1996, Newt Gingrich and many in my class tried to end such a regime of central planning with the Freedom to Farm Act. We gave farmers more freedom to choose what they would plant and cut back on government interference while setting up a system that would wean corporate farmers from government subsidies by 2002. Unfortunately, when 2002 rolled around, Congress lost its political will and abruptly reversed course.

The reason? The most competitive Senate and House elections were in farming and ranching states. Republicans and Democrats were therefore willing to spend billions of dollars to win races in those farm states. By paying out record sums in farm subsidies, party leaders made the pitch that their candidate should receive your vote because of all his party had done for the farmer—even if agricultural corporations got most of the money. The Bush White House went along because many of the "red states" that went for Bush 43 in 2000 were those same farm states Bush would need to win re-election.

The calculation was simple: lose one farm state in 2004 and you lose the election. Thus began the drafting and passing of the largest corporate-welfare scam in American history. But to pass this legislative bribe through Congress, senators and congressmen would have to be pulled into the conspiracy.

Two billion dollars was drummed up to support dairy farmers. This brought in the Maine delegation, Pennsylvania senator Arlen Specter, and Vermont senator James Jeffords. Another $4.4 billion for peanut farmers bought the Virginia, Alabama, and Georgia delegations. And since sugar is king in Florida, Sunshine State members climbed on board the farm-subsidy gravy train. Similar packages for cotton, wheat, wool, mohair, and even apples and onions were sprinkled in. There were also smaller handouts: like the $631,000 for "alternative salmon products" in Alaska.

Standing alone, not one of these corporate welfare measures could survive the bright light of public scrutiny. But the peanut states vote with the dairy states so that the dairy states will vote with the peanut states. Congressmen with mohair in their district back congressmen with wheat in theirs. In exchange the wheat member becomes a fierce defender of the mohair subsidy. And so on down the line. As you see, strength grows in numbers.

Once all the deals were cut and all the reciprocal back scratching was done, the price tag for the most recent farm bill jumped up to a staggering $190 billion. This was an $83 billion increase over the already ridiculous program in place—the same program Congress promised to abolish in 1996. Once again the Washington status quo beat back well-intentioned reform.

Sadly, the farm bill is not a unique case. Congressmen and senators swap support for pet projects all the time. Every time a member wants to spend one dime of your tax money, he will end up spending a stack of your bills in order to buy the requisite votes needed to pass his bill through Congress.

Herding Fat Cats

So why can't a maverick who wants to end the trillion-dollar back-scratching scheme in Congress survive in Washington? Let's take it step by step.

We will imagine that this book has so inspired you (work with me here) that you want to follow in my footsteps and try to change the way Washington works. So you make the same decision I made ten years ago and launch a long-shot campaign for Congress. Like me, you will probably start this journey with no money, no name recognition, and virtually no hope of winning.

Perhaps it's no coincidence that even the fictional Mr. Smith was appointed and didn't have to unseat an incumbent. In the last election, only four outsiders in the entire country defeated a sitting member of the House of Representatives. That means the other 431 breezed to re-election. Those numbers highlight just how sick our political system is, especially when you consider that the old Soviet Union had a higher turnover in their Politburo than America has in Congress.

Let's carry this forward and assume that lightning strikes. You knock on every door in your district. You inspire your neighbors. You marry a wealthy heir or heiress to bankroll your campaign. Somehow, you get elected.

Congratulations, Congressman! Now the real fun begins.

As soon as you step off the plane in D.C. you will immediately notice two things: First, the weather is awful. Second, the party leadership owns your political life.

So you want to clean up government? That's great, boy, but you'll need a seat on the Government Reform Committee.

Now, how in the world would a young political buck like yourself ever manage to get on a congressional committee like that? Take a test? Enter a debating contest with other new members? No. This ain't no meritocracy. You will only get your assignment by kissing up to the Speaker of the House and a handful of the most powerful committee chairmen on Capitol Hill. Oh, yeah. You probably figured this out by now, but these D.C. power brokers didn't earn their horns by trying to reform anything. If you want your committee assignment, it's time to bow and scrape. And once you get it, you will owe them big.

Try to imagine the trajectory of your political career the first time you decide to follow your heart and try to kill one of your party's god-awful, pork-filled pieces of crap. Instead of *Mr. Smith Goes to Washington,* it will play more like *Kill Bill, Part III*. By the way, tonight you will be playing the role of Bill and it will be a very bloody ending. Unlike Uma, you probably won't survive.

But let us assume that your goals are a bit more modest. Maybe you just want to talk about cleaning up government. Get some press. Shine a spotlight on an issue. Well, in order to get floor time during a debate that actually means something, you will have to talk to leadership or the powerful committee chairman who is running the business on the floor that day.

Good luck, Maverick.

And, of course, if you need money for anything—new roads in your district, research funding at the local university—you'll most definitely have to talk to the leadership. Congress is a small fraternity, and if you cross one member of that club by killing his pet project, swarms of his fellow frat members will crush any bill you propose in the future regardless of its merit.

In this country club memories are long and payback is hell.

"Wait!" you say. You're a reformer who's running to end pork-barrel spending. You don't need money for your district. You don't need a seat on the Government Reform Committee. You're going to rally the public to demand change. You don't even need to speak on the House floor. You'll stand on the Capitol lawn and hold press conferences.

Fine, but you better not unpack those boxes you just shipped up from Dogpatch, Florida, because your crusade will last a total of one term (that's two years for humans)—an inconsequential blink of the eye.

But again, for the sake of argument, let's assume the people in your district have infinite patience and are willing to drive over potholes, watch that new monorail bypass their town, watch the university close down labs, and allow themselves to be comforted by the fact that their neighbors in the adjoining district are enjoying all the fruits of your district tax dollars.

You still have a problem. Someone is going to run against you and in order to defeat him or her, you're going to need money. Lightning may sometimes strike, but rarely does it strike twice in the same place in the world of politics. You may have been like Mr. Smith or the long-shot Rocky Balboa the first time you ran for Congress, but this time you are the political insider, even if other members of your own party are working overtime to help your challenger. And, yes, Cowboy, they will do that to help quietly unseat a troublemaker like yourself.

It's easy to get bogged down in all the complex regulations that govern political fundraising, but let me simplify the realities of campaign financing in the twenty-first century for you and other potential candidates.

You need to raise and spend a sickening amount of money to get elected and re-elected. And since heiresses like Teresa Heinz Kerry are prone to mood swings, don't count on spousal financing after your first election. This is especially true because your wife will learn, like most spouses, that being a congressman's better half ain't all that it seems.

Anecdotal evidence aside, the hard numbers provide even less comfort. During the 2002 election cycle, the candidate who spent the most money won 90 percent of the time. And spending the most money means raising the most money. Buying a seat in the United States House of Representatives cost roughly $900,000 the last election cycle.

So how do you raise all that money? You start by putting together a Rolodex of fat cats who need something from you, your committee, or your party leaders. If you have spent your first term in Congress being a political maverick and thumbing your nose at political bosses, you ain't going to get jack from your committee chairman or party leaders who see no reason to reward independent spirits.

This is a big problem. While you may be charming, most people outside your district aren't willing to pay $1,000 to eat cold chicken in a hotel ballroom and listen to you speak. However, as baffling as this may seem to those who do not work inside the Beltway, people will line up around the block and pay $1,000 to listen to spellbinding speeches from the likes of Speaker Denny Hastert and Tom Delay. Mr. Delay is a national figure. Leader Delay can herd the fat cats into the ballroom of your hometown Best Western as well as any other place.

Unlike you, Tom Delay can make things happen in Washington. Unlike you, Tom Delay is universally loved by fat cats. This is probably because the majority leader talks to them, dines with them, and golfs with them. You on the other hand would not be allowed on the same golf course with them because they'd be afraid to turn their back on you, for fear of where you may stick your nine iron.

Only party leaders can drum up the cash you need to keep a seat in Congress, whether through direct contributions to your campaign or through political action committees (PACs). And yes, nimrod, they are going to want something in return. That something is loyalty—especially when it comes to appropriations votes. And as we've seen, party leaders are career politicians beholden to all the malfunctioning institutions that hemorrhage tax dollars. Frugality is the last thing that most leaders expect from you when it comes to appropriations votes.

Admittedly, a rare few do survive outside the good graces of their parties. Senator John McCain is the best example, but he had already served several terms in the Senate and built a solid political organization before becoming a reformer. Since making that move he has had to rely on contributors from all over the country to keep his campaign coffers full. Eighty-six percent of the money he raises comes from outside his own state, and a lot of it comes from Independents and Democrats drawn to his reform message. Still, at times Senator McCain has been reduced to delivering floor speeches to empty Senate chambers and issuing blistering press releases to the faithful that make up his huge national following—while the rest of his colleagues continue to spend tax dollars like drunken sailors.

Yes, *Mr. Smith Goes to Washington* has a happy ending. But the endings in Washington seldom are as sweet. Instead, most members stumble to a forced exit that is preceded by years of frustration, humiliation, and compromise.

In the end, one man doesn't have a prayer when faced with a system so predisposed to waste taxpayers' money. What is needed is more than you playing the lone horseman of the apocalypse. What is needed is a cavalry.

Chapter Ten

How to Fix It

IT'S EASY TO HIGHLIGHT the many ways Washington screws things up. Late-night talk show hosts do it, cable news anchors do it, and newspaper pundits do it. Even Eddie Murphy made a movie about it. Washington's corruption, inefficiency, and just plain stupidity make for a pretty broad target.

Complaining about how Washington spends our money is a lot like complaining about how the Red Sox can't win a World Series. Some things seem to be immovable certainties of the Universe. However, while the ineptitude of the Sox only tortures those of us in Red Sox Nation, the habit Congress has for hemorrhaging tax dollars punishes all Americans. More important, it poses a real threat to the long-term health of our democracy.

So what's the solution?

Placebos

The most casual political observer understands that the intersection of campaign money and politics is the root of the problems in Washington. Recent efforts by Senators John McCain and Russell Feingold to reform the campaign finance system have received much attention and praise from media elites. I suppose the easiest way for me to conclude this book would be to tell you that their campaign finance bill will cure all that ails Congress, clean up the corrupt system, and allow us all to sleep peacefully tonight. Unfortunately, that is not true.

For those of you who don't read *Roll Call* all the time, here's an update on where campaign finance rules currently stand. President Bush signed the Bipartisan Campaign Reform Act (known as the McCain-Feingold bill) into law on March 27, 2002. Immediately afterward, Republican senator Mitch McConnell joined the Republican Party and special-interest groups challenging the law in court. Several of its provisions looked as if they would be overturned, but this past December the Supreme Court ruled 5 to 4 to leave the bill largely intact. As a result, the 2004 election will be the first governed by its restrictions.

The biggest change made by the McCain-Feingold bill is

its ban on so-called soft money. Those are the unlimited contributions individuals, corporations, and labor unions are able to make to the two political parties. According to the campaign watchdog group Common Cause, the Republican and Democratic Parties raised nearly $500 million in soft money during 2000 and 2002. That money helped big business and the unions keep their grip on the leadership of both parties. Eliminating it *may* help reduce such influence.

I say *may* because the big contributors are already finding ways around the law. So far the most popular method of cheating is through organizations called "527s," named for the section of the tax code they fall under. Instead of going to either Democrats or Republicans, money is now pouring into these 527s with millions of dollars being spent toward partisan efforts whose goal, just like the parties, is to elect one candidate or another. The FEC could crack down on the 527s, but you can be sure that big money will find another hole in the reform net.

Further, by eliminating soft money from corporations and unions, McCain-Feingold forces the parties to rely more on wealthy individuals. Because the bill also raises the amount a wealthy individual can give, the parties now have greater incentive to court more rich people. That means politicians now will spend more time herding fat cats. It's safe to assume most people who have thousands of dollars to give a political party didn't earn that money running homeless shelters. They probably made it in a business—and probably one that could benefit from Congress voting one way or another on laws and regulations. In short, the influence-buying game remains alive and well. Anyone who doubts this need only read reports of how Democratss used their 2004 convention in Boston to herd the fattest and richest of cats.

All of this is not to say that campaign finance reform is not a noble goal. It is to say we shouldn't place our hope in attractive placebos that don't offer lasting cures.

Get a Real Job

If we accept the influence of big money will be hard to eliminate, we must look for ways to blunt its impact. First, Congress can definitely make it harder for money to buy influence by changing the rules about lobbying.

Members of Congress and Hill staffers should be barred from lobbying for five years after they leave their government jobs. Too many go right to lobbying firms where they use their old connections and stature to distort the business of government. Even though there is a brief ban on ex-members lobbying, it is so short that retired or defeated congressmen are snatched up by lobbying firms while they wait out this one-year ban. During this time former members still find ways to sell direct access to the decision makers who are supposed to be listening to the people.

The problem is that if you're the former Speaker of the House turned lobbyist, chances are that a group of citizen activists can't afford to retain your services. Only big wealthy special interests can. So I say let's level the playing field. Send the former Speaker home for a few years to get a real job. Make it harder for the special interests to buy easy access. Let them set up a phone bank and try to mobilize voters instead. If nothing else, it would be amusing to watch the CEO of a pharmaceutical company calling senior citizens and asking them to write their senator about the new prescription drug bill.

Truth in Advertising

Another way to blunt money's impact is through "real time" campaign reporting. Currently candidates running for federal office are required to report the names of their contributors on a quarterly basis. This makes it hard for the public to see how money affects the decisions their representatives make. For instance, let's say Congress is about to vote on a new multibillion-dollar subsidy for mohair farmers. Let's also say that a week before the vote the mohair industry starts giving big money to certain congressmen and senators. Coincidentally, those congressmen and senators suddenly become very vocal supporters of mohair.

When your local newspaper covers the story of the Great Mohair Debate, they'll be able to tell you a few things. They'll be able to tell you the mohair subsidy passed. They'll be able to tell you that your congressmen and your senator voted for it. And they'll even be able to tell you that your senator gave a rousing speech that helped the pro-mohair forces win their victory.

But the newspaper won't be able to tell you what mohair is. (To be honest, I'm not entirely sure.) And they won't be able to tell you that the mohair industry started funneling big bucks into your senator's and congressman's campaign coffers just days before the vote. If you're thinking that little detail is the most important part of the whole story, you're right. And, that's where "real time" reporting comes in.

As the system works now, the mohair industry's contributions will be reported, but it could be months later and certainly long after you and your local newspaper have turned your attention to something else. If we had a real-time system, those contributions would be posted on the Internet as soon as they were made. And before any big vote, you'd be able to see who was pumping cash into Washington. Journalists could tell you that critical part of the story any time they reported on a big decision in Congress.

The technology exists to make this all possible. Real-time reporting is already mandated for state-level races in Michigan. And during the 2000 race, President Bush's campaign voluntarily posted all donations as they were processed.

A system of real-time information brings transparency to the political process and hopefully a little bit of shame. Congress will be less likely to cash in on industry donations before big votes if they know their constituents can read about it the next morning.

Six and Out

The case for real-time reporting is a good reminder of one fundamental fact that often gets forgotten in so many crusades for reform: At the end of the day, voters have all the power. That's a Civics 101 lesson, but it usually gets lost in all the fancy proposals for more regulation. If you don't like the way Washington is spending your money, vote for candidates who will spend it differently.

I know what you're thinking: *Given all the forces conspiring against him, how can I be sure my reformer candidate will stick to his word when he gets to Washington?* Well, I do have good news. There is an antidote to the D.C. disease, an armor that will keep the Fat White Pink Boys at bay. It's called term limits. Candidates without a track record of reform like John McCain should agree to serve a short period of time—three terms in the House, for instance—and then return to private life. If they don't make the promise, then they don't get your vote.

Candidates who agree to term limits can avoid the temptations of deal making that trap those looking to stay in Washington for the long haul. Those candidates are much less likely to say to themselves, "Well, I guess I can violate my principles and vote for this new highway now so maybe I can get more done next term." Rather, they need to raise less money to remain independent of big business, union bosses, and party leaders. This makes them far harder to threaten with the traditional arsenal of Hill repercussions.

I remember South Carolina governor Mark Sanford and I being berated by congressional leaders when we both were serving in the House. One leader warned of dire consequences if we didn't cooperate. Because Sanford was self-term limited and I was half-crazy, we both broke out laughing after the threat was made. Sanford looked at me and said in between his chuckles, "Did they just threaten us?"

Obviously the threat had no impact on Sanford or Matt

Salmon (R-Ariz.) or Tom Coburn (R-Okla.) or my other 1994 classmates who pledged to serve three terms, then leave.

Evidence of the positive effects of term limits is far more than anecdotal. Several organizations have studied the phenomenon of self-imposed term limits and all came to the same conclusion: Candidates who have agreed to leave Washington after a set number of terms have a proven track record of fighting spending.

The best data comes from the nonpartisan National Taxpayers Union Foundation (NTUF). They looked at the Freshman Class of 1994 and tracked our attitudes toward spending. They found that those few who had committed to limit their terms consistently voted to reduce spending throughout their tenure. In contrast, the majority who did not self-limit started advocating spending increases by the time their second term rolled around.

The evidence gets even more compelling. The NTUF found that the Class of 1994 actually advocated more and more spending for every term they spent in Washington. The term-limiters, on the other hand, pushed for more and more cuts the closer they got to their self-imposed retirement.

There have been other studies as well. The CATO Institute found that the longer a member served in Congress, the less likely he or she was to support spending cuts. Examining the 104th and 105th Congress, CATO discovered junior Republicans were "more than twice as likely to vote for spending or tax cuts as were senior Republicans."

On top of these numbers, my own personal experience made me a real believer in limits. As explained earlier, Matt Salmon, Mark Sanford, and Tom Coburn all had committed to serving only three terms. I saw how that commitment gave them the independence to challenge Newt Gingrich when he wavered on tax and spending cuts and allowed them to stand up to the leadership on so many wasteful spending proposals.

More recently while watching the debate on the Farm Bill, imagine my surprise when former Texas senator Phil Gramm stood up to oppose the measure's profligate spending despite the huge windfall it would bring his own state. I can assure you that Mr. Gramm did not wake up in the middle of the night overcome with guilt at the prospect of bankrupting his grandchildren with a massive federal debt. The conservative Texan had simply decided to retire and was finally free to make all of his decisions based on what was right and not what would win elections.

How to Reform Washington

I've already said in order to really change Washington, it's going to take more than a few committed mavericks. We need another revolutionary guard like the band of barbarians who stormed the Imperial Congress in 1994. And once again, we must elect candidates who will focus on protecting taxpayers and clean up Washington.

In 1994, Newt's Republicans all famously signed the Contract with America, a promise to change the way Congress did business through essential reforms. We accomplished a great deal those first few months, but as I've described in this book, the promises of reform were never fully realized.

Frank Luntz, the pollster who wrote the first Contract with America, has devised a new set of reforms he calls the "Taxpayers Bill of Rights." It's a series of measures that would simplify how Washington collects your money and restrain how they spend it. I have borrowed from Luntz's proposal and combined his ideas with my political reforms; I believe these ten proposals would change the way Washington works:

1. Ban congressmen, senators, and White House officials from lobbying for five years.
2. Freeze the pay of congressmen, senators, and White House officials until the federal budget is balanced. *This includes cost-of-living adjustments!*
3. Force political candidates to immediately scan and post all campaign contributions on their campaign website. Failure to do so results in criminal penalties.
4. Pass term limits now! Since the House of Representatives authorizes the federal spending, limit House members to three terms (six years).
5. Make Congress and every Washington bureaucracy undergo an independent, professional audit, line by line, program by program, every four years.
6. Pass a constitutional amendment requiring Washington to balance the budget every year except when Congress passes a resolution declaring a national emergency.
7. Create a federal rainy-day fund that would set aside one-half of one percent of all tax receipts each year for national, state, and local emergencies.

8. Reenact pay-as-you-go rules that would require Congress to offset new spending programs and tax cuts with spending cuts from other programs.
9. Reinstitute congressional spending caps that would force congressmen or senators to live within their previous spending projections. These caps will not be broken unless Congress passed a separate resolution declaring a national emergency as described in number 6.
10. Pass a new American tax code written by a bipartisan panel of budget experts instead of the lobbyist groups who regularly carve out special-interest deductions and greatly simplify the tax system.

This year on *Scarborough Country* I will invite incumbents and challengers of both parties to sign my 10-Point Pledge to clean up Congress. I'll be highlighting candidates who agree to support all provisions of the reform package and demanding answers from those who don't. If enough voters in enough districts in enough states demand accountability from their representatives, we can take back our government and send corrupted incumbents and the Fat White Pink Boys packing.

When I first ran for Congress years ago, I did so with the idealism of a twenty-nine-year-old who thought hard work and good ideas could change the way Washington works. Perhaps I have grown more cynical than the young candidate who shocked friends and foes alike with a historic win. But today, I still believe in the greatness of America. Rome may be burning, but the American people have proven time and again that they are ready to grab a bucket and join the fire brigade. This country has an extraordinary tradition of peaceful citizen revolt, and such revolts have fed our vital tradition of reform.

These past few years have been a time of great crisis for America and the world. Matters of national security have to trump all other concerns. But we cannot let our government drift under the spell of a spending frenzy, fueled by opportunists and profiteers. Every two years we have a chance to correct course and another such chance is just around the corner.

Somewhere out there another thirty-year-old fool is thinking maybe he can win a seat in Congress. He's thinking maybe he can change the way Washington runs its business. He believes that he and a band of brothers can fight the special interests, cut spending, shut out the lobbyists, and give the people a new voice. Maybe he'll succeed or years later return home a defeated man. But tonight, with so much at stake in the country I love, I pray to God that fool runs.

Epilogue

In politics as in life, timing is everything. My first day practicing law after retiring from Congress was to be September 11, 2001. I dropped my son Joey off at school and swung by the *Florida Sun* office to grab a copy of the latest edition. Patty Lowery, who was also returning from dropping her son at school, stopped in the middle of a busy intersection and rolled down her window.

"A plane just flew into the World Trade Center!" she shouted.

I jumped in the car and turned on our local AM talk station WCOA. By then, everyone knew America was under attack. In minutes the world had changed forever.

I spent the next few weeks grieving with the rest of America over the losses we endured that day. At night I would walk around my home aimlessly wondering how it was possible that my first work day out of Congress in seven years was the very day America was attacked by these terrorists. There was little I could do that a united Congress and president couldn't without me; but being a thousand miles away from my congressional office during this epic crisis didn't seem right.

Yet it was.

God only knows that God has his plans, so Paul Simon says. And just as I believe God had a hand in my decision to run for Congress, I know my departure was part of a better plan. When I announced my retirement, I said that at the end of my life I would rather be remembered as a good father than a great congressman. The crowd applauded, party leaders smiled and whispered amen while staff members came up and congratulated me for having my priorities straight.

Those same staff members had seen my seven- and four-year-old boys grow up to be fourteen- and eleven-year-old young men. The seven years of separation from my loved ones were especially tough because the Republican Revolution was far from family friendly. Most sessions kept me away from my family more than two hundred nights a year, while the ever-raging budget battles between Clinton and Congress seemed to drag on longer each year.

The strain was especially tough on my oldest son, Joey. He was a fourteen-year-old who loved his mother but had put up with a part-time dad for too long. I had always told him that whenever he needed to talk to me I would be there. If that meant calling my cell phone while I was in the middle of a meeting with constituents, party leaders, or presidents, I would leave that meeting to talk to him because he and his brother always came first. They were reassuring words for a while, but middle-school boys can't be raised by cell phones. His mother and I had separated in 1997, and by the time he was a teenager, everyone in our family knew that either Joey needed to move to Washington to live with me or I needed to move back to Florida to be with him. Because his mother and younger brother, Andrew, were in Florida, it wasn't even a close call. Andrew was younger and had different needs than Joey. He had recently been diagnosed with type 1 diabetes. As any parent who has a child with diabetes knows, the disease places a great many challenges on family members. Almost all of those burdens fell squarely on the shoulders of Andrew's mother while I served in Congress. Ultimately, it was time for me to pull my weight as the boys' father and leave my congressional career behind.

It's hard to tell you what an honor it was for me to be sent to Congress by the voters of Northwest Florida. Despite run-ins with Republican leaders and the Democratic president, Congress was the most rewarding job I could ever have. I began working at seven in the morning and usually finished by nine or ten o'clock at night. I didn't do receptions, stayed away from golf courses, and never learned anything about Washington nightlife.

I spent almost all of my time working behind my desk in Congress because I loved my job and was determined not to let down those who took a risk on a young, unknown political novice.

My love for Joey and Andrew and my dedication to the

work is why I still say quitting Congress was the easiest decision I ever made—and the hardest. When I first arrived on Capitol Hill, older members warned me that I needed to keep my children close. Several senior members told me similar stories about how Congress destroyed their relationship with a son or daughter. Two of these old men pointed to pictures of their thirty-something sons or daughters and started to tear up. They had lost their children because of politics, and I was warned not to make the same mistake. Maybe that's why the only people who weren't shocked by my early retirement were other members of Congress.

I was always considered to be a young man in a hurry when I was first elected to Congress. That may have been why my abrupt departure from politics left reporters and conspiracy theorists grasping to explain why a thirty-eight-year-old man would give up a promising political career. I told reporters about coming home for my boys but felt no need to go into every last detail. They didn't buy the truth. Local newspaper staffs began whispering to themselves, and others, that I must have gotten someone pregnant. The editor of Ft. Walton's *Northwest Florida Daily News* picked up on this lie and wrote in a column to his readers that I would be a proud papa within months. When I called him up to ask if he knew something I didn't, the editor sheepishly admitted he had heard the rumor in his newsroom and ran with it.

I was shocked.

A few months after announcing my retirement an office worker died while working in a satellite office in Okaloosa County. I had met the young woman only a few times at random public events, but I sincerely felt sadness for her family.

Liberal conspiracy theorists on the Internet quickly seized on this young woman's death to try and explain my retirement from Congress. The rumor now being spread was that I quit Congress because I got a staff member pregnant and killed her. It was a disgusting lie that added a great deal of pain to the woman's parents and husband who now had to endure the heartache of her passing. It was despicable and made me glad I was leaving politics.

Other rumors involving drugs, bribes, and barnyard animals greeted my departure from Congress as well. A high school student told me months later that he and his father had always looked up to me, but now his dad told him I left Congress to make lots of money. I felt like a modern-day Shoeless Joe Jackson reassuring a kid who seemed to be saying, "Say it ain't so, Joe." It wasn't. I gave up hundreds of thousands of dollars by deciding to work in Pensacola, Florida, as a former congressman instead of Washington or New York. But I didn't tell the kid. By this time, I had learned that there was no reason to try and disprove conspiracy theories. God, my family, and my friends knew the reasons I was leaving, and that was all that mattered. But as someone who worked around the clock for seven years, won every election in a landslide, and generally steered clear of negative press coverage, this nasty little tempest was a bad ending to a wonderful political career.

I thought it interesting that other members who quit Congress to run for the Senate or a governorship got accolades from the press and fellow party leaders. But no one, including many of the so-called family-values groups, stepped forward to suggest that it was acceptable to retire from Congress for your children's well-being.

My sendoff was sloppy, but in the end I really didn't give a damn. I got into Congress against all intelligent advice. And seven years later I got out the same way.

I have never looked back. MSNBC producer Peter McCarthy once asked me if I missed Congress. I immediately shot back, "Do you miss high school?" The answer was no. Part of this has to do with the nasty nature of the business, but part of me was relieved to be out of the fight for a while. I knew if I had stayed in Washington I would not be fighting Bill Clinton and Ted Kennedy, but George W. Bush and Denny Hastert. Spending was spiraling out of control and it was my party that was playing fast and loose with the money. After I retired, former constituents seemed let down when I answered no to their question of whether I missed Congress. But I was doing things for the first time that others took for granted. That's why coaching baseball, driving the boys to school, watching football games with them on lazy weekends, and doing things that regular dads do was so much more rewarding to me than meeting presidents and prime ministers.

For some reason, voters in my home district always called me "Congressman Joe," "Regular Joe," or just "Joe." Maybe that's because I refused to answer to "Congressman Scarborough." But there was nothing regular about my life in Congress. Professionally, that was a plus. As a father, I thought it was a huge minus. As shocking as it may seem to the Fat White Pink Boys, by the time Joey and Andrew reached middle school, my paternal calling as father drowned out all other ambitions.

Soon after leaving Congress, I called John Shadegg (R-Ariz.) to ask what it was like having a Republican president to complement Congress. John said, "We've torn down in seven months what it took us seven years to build." Any good news on spending issues was more likely to come from the Bush administration that was set up in Tallahassee, Florida. The president's brother Jeb declared war on government spending and tax increases as soon as he entered the Florida governor's mansion. I spent the first few months of Jeb's first term taking angry phone calls from supporters who wanted to know why the new governor was vetoing their favorite state-funded project.

"Maybe because that's what he promised to do," I'd reply.

Jeb Bush had no problem calling the pet projects of his biggest supporters "turkeys" while lining them out of spending bills. At times, it seems, he went out of his way to aggravate allies with his veto pen. But maybe that's because so few seem to possess political courage these days.

Another rising political star is Mark Sanford, who left Congress in 2000 to run for governor of South Carolina. Sanford and I were together on most fights against our leadership when they backed down on spending cuts or reform measures. Unlike most politicians I know, Mark Sanford has only become more courageous with his political promotion over the years.

This year the Republican legislature in South Carolina went on a spending spree of record proportions. So Governor Sanford vetoed more than one hundred spending bills in one day. The Republican House and Senate overturned most of those vetoes the next day, prompting a response from their Republican governor. Sanford, who shares my lack of interest in impressing the political establishment, decided to carry two pigs to the South Carolina House chambers and confront his big-spending Republican legislators. Sanford named the pigs "Pork" and "Barrel," and paraded them around the GOP House for all the members to see. They were, of course, greatly offended—perhaps it was the droppings left on the House floor courtesy of Pork and Barrel—but Sanford couldn't have cared less. He was more interested in being faithful to South Carolina taxpayers than to a political party. If there were more Mark Sanfords and Jeb Bushes in politics today, we would not have a $7 trillion national debt.

Many Republicans and self-described Democrats remind me of a musician who once played in my band who told me of the story of how his girlfriend came home early to her apartment to find him locked in the arms of another woman.

"What did you do?" I asked.

"I did the only thing I could do. I jumped up and started yelling at her, saying if she couldn't trust me enough to leave me alone once in a while, our relationship would never work out."

Remarkably enough, she bought it.

But you don't have to. Your elected Washington representative is part of a Congress that has wasted more tax dollars this year than ever before. When you see him or her on the campaign trail, ask why the national debt is approaching $7.5 trillion. Ask why the yearly federal deficit is at a record high. Ask how Americans are going to pay for the trillion-dollar Medicare drug plan, or how taxpayers are going to cope with the financial meltdown facing America when baby boomers start drawing down the Social Security and Medicare trust funds.

Chances are he will have not a single good answer—just a

lot of political babble. Republicans will blame Democrats. Democrats will blame Republicans. No one will give you a straight answer. But don't be baffled by their B.S. You have caught your representative in the arms of your tax dollars. He hasn't given you a single reason to trust him. He is part of an institution that has taken a yearly $155 billion surplus and turned it into an annual $455 billion deficit. And your elected official is part of a system that has put the future of our country in peril because Washington politicians are more concerned about their political future than protecting the future of you and your family.

The passing of Ronald Reagan may have coincided with the death of the Republican Party's soul as the *Wall Street Journal* suggested, but going back to Washington for Reagan's funeral did remind me of all the reasons I got into politics. It wasn't because of Newt Gingrich or the Contract with America. It certainly wasn't because I wanted to blindly follow a political party over the cliff. It was because I grew up reading books about American leaders who believed that one person could make a difference in this great country of ours. Ronald Reagan showed how one person's courage could help free the world from Soviet tyranny, and how that victory could free an entire continent.

Twenty years earlier it was Bobby Kennedy who spoke of how one person could change the course of history and bring down the mightiest walls of oppression. Kennedy spoke those words in South Africa, but a few years later he was the lone voice making the difference on the night Martin Luther King died. RFK was in Indianapolis that evening and wanted to deliver a message of condolence to those who saw King as their last great hope. Police officers strongly warned him against going into the inner city to deliver news of the assassination. Police even refused to provide Kennedy with protection because riots were already breaking out on the East Coast. But the senator ignored these warnings, delivered his speech, and urged the residents of Indianapolis to pay tribute to the life of Martin Luther King by rejecting those calling for violence. As Arthur Schlessinger Jr. wrote in his book on Bobby Kennedy's life, even though riots broke out that night in Chicago, New York, Los Angeles, and scores of other cities, Indianapolis went to sleep in peace.

Bobby Kennedy said few possessed the greatness to bend history by themselves, but together we could work to change the world. While in Congress, I saw men and women work together to make America better, stronger, and safer. Leaders like Matt Salmon, John Shadegg, John McCain, Steve Largent, Mark Sanford, Tom Coburn, and Steve Chabot more often than not dared to challenge the established order of Washington. There certainly were, and still are, others. And I know there are even a handful of Democrats concerned about the exploding federal deficit who will cast votes to match their rhetoric.

But Washington politicians won't do it alone. You have a duty to hold all your elected representatives accountable and to let them know that you are watching their votes. And if they don't shape up, you will do all you can to send them packing.

And if you decide to run for a big-spender's congressional seat, give me a call. I think it's about time for another peaceful citizens' revolt.